REVISE EDEXCEL GCSE
Science
Extension Units
REVISION WORKBOOK

D1428697

Series Consultant: Harry Smith
Series Editor: Penny Johnson

Authors: Peter Ellis, Damian Riddle and
Stephen Winrow-Campbell

THE REVISE EDEXCEL SERIES
Available in print or online

Online editions for all titles in the Revise Edexcel series are available Autumn 2012.

Presented on our ActiveLearn platform, you can view the full book and customise it by adding notes, comments and weblinks.

Print editions

Science Extension Units Revision Workbook	9781446902585
Science Extension Units Revision Guide Higher	9781446902677

Online editions

Science Extension Units Revision Workbook	9781446904671
Science Extension Units Revision Guide Higher	9781446904664

Print and online editions are also available for Science (Higher and Foundation) and Additional Science (Higher and Foundation).

This Revision Workbook is designed to complement your classroom and home learning, and to help prepare you for the exam. It does not include all the content and skills needed for the complete course. It is designed to work in combination with Edexcel's main GCSE Science 2011 Series.

To find out more visit:
www.pearsonschools.co.uk/edexcelgcsesciencerevision

ALWAYS LEARNING

PEARSON

Contents

A small bit of small print

Edexcel publishes Sample Assessment Material and the Specification on its website. This is the official content and this book should be used in conjunction with it. The questions in this book have been written to help you practise what you have learned in your revision. Remember: the real exam questions may not look like this.

Target grade ranges

Target grade ranges are quoted in this book for some of the questions. Students targeting this grade range should be aiming to get most of the marks available. Students targeting a higher grade should be aiming to get all of the marks available.

Rhythms

D-B 1 Five plants were exposed to different lengths of day and night. After ten weeks it was noted whether flowers had formed on the plants. The results are shown in the diagram below.

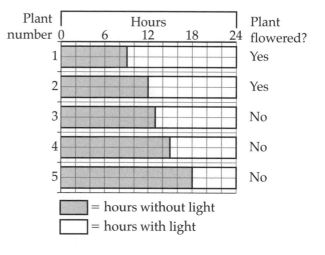

a) State the minimum number of hours of light needed to make the plant flower, giving a reason for your answer.

The minimum number of hours needed to make the plant flower

is .... because

....

(2 marks)

> On every page you will find a guided question. Guided questions have part of the answer filled in for you to show you how best to answer them.

Guided

b) The five plants were from the same species. Suggest **two** other factors that should be kept constant so that the results can be compared.

Plants should be..identical and there should be

the same...intensity each time.

(2 marks)

c) Suggest **two** reasons why plants of the same species flower at the same time.

....

....

(2 marks)

D-C 2 a) State what a circadian rhythm is.

....

(1 mark)

b) Circadian rhythms in humans often involve hormones being released. For example, a hormone called melatonin is usually secreted into the blood stream about 10pm each night. Melatonin makes us feel sleepy. Suggest what might happen to the level of melatonin during the rest of the 24-hour period.

....

....

....

(3 marks)

c) Explain why flying east or west can cause people to feel sleepy during the day. Use what you have learnt about melatonin in the question above to help you.

....

....

(2 marks)

Plant defences

D-C

1 A student investigated whether garlic contains a chemical that kills pathogens. She crushed some garlic in a little water to make a juice. She mixed this juice in a sterile tube with a bacterial culture and a nutrient-rich jelly. She then put a lid on the top of the tube and left it for 2 days.

> This question is about a practical piece of work. It is a good idea to look at your practical work before the exam just to remind yourself of the skills you use when carrying out an experiment.

She then repeated the investigation but used water rather than garlic juice.

a) What is a pathogen?

....

(1 mark)

b) What is the purpose of the investigation that used water only?

....

(1 mark)

c) State the name of the part of the investigation that used water only.

....

(1 mark)

> Guided

d) Suggest **two** reasons why a lid was placed over the top of the tubes.

A lid was placed over the top of the tubes to stop other bacteria from getting in.

....

(2 marks)

e) State **two** factors, other than those mentioned in the question, that should be kept constant in this investigation.

....

....

(2 marks)

f) State an advantage to the garlic of containing an antibacterial substance.

....

(1 mark)

C-A*

Higher

2 a) State **two** different defences plants may use to try and reduce caterpillar numbers.

....

....

(2 marks)

b) Explain how caterpillars can reduce the yield from a crop.

....

....

....

(3 marks)

Growing microorganisms

1   A population of bacteria doubles in number every 20 minutes. A sample contains 10 bacteria at 0 minutes.

a)   Calculate how many bacteria there are after 1 hour (60 minutes).

After 20 minutes there are _____ bacteria.

After 40 minutes there are _____ bacteria.

After 60 minutes there are _____ bacteria.

> The rate of population growth will not continue like this because something such as nutrient supply , will start to limit it.

(1 mark)

b)   State the name for the type of growth where numbers double in a set time.

....

(1 mark)

E-B   **2**   Four cartons of pasteurised milk were kept in different conditions. The conditions are shown in the diagram below.

Milk cartons on kitchen work top   Milk cartons in fridge

   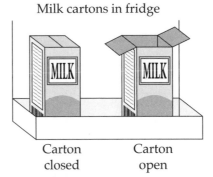

Carton   Carton   Carton   Carton
closed   open   closed   open

a)   Describe how the milk has been pasteurised.

....

....

(2 marks)

b)   Eventually all four cartons of milk will go off. What does this tell you about the milk in the cartons?

....

....

(2 marks)

c)   Explain which carton of milk is likely to go off first. Give reasons for your answer.

....

....

....

(3 marks)

Had a go ☐ **Nearly there** ☐ **Nailed it!** ☐

Vaccines

F-C 1 The table below shows three situations and the likely outcome for the person involved.

Situation	Likely outcome for the person involved
1 Person exposed to cowpox virus	caught cowpox
2 Person exposed to smallpox virus	caught smallpox and died
3 Person exposed to cowpox virus and then later smallpox virus	

Guided   a) Complete the rows of the table. **(3 marks)**

b) Explain why the person in **Situation 1** caught cowpox.

....

....

(2 marks)

c) Explain what is likely to happen to someone who has been exposed to cowpox and then smallpox.

....

....

....

....

(4 marks)

> Make sure you know the difference between antigens, antibodies and antibiotics.

C-A* 2 a) Describe where antigens are normally found.

Higher

....

....

(2 marks)

b) Suggest why a person might be injected with an antigen.

....

....

....

(3 marks)

Antibodies

Higher This whole page covers Higher material.

C-B 1 a) State what causes memory lymphocytes to first form.

....

(1 mark)

b) State what causes the secondary response to occur.

....

....

....

(2 marks)

c) Describe the role of memory lymphocytes in the secondary response.

....

....

....

(3 marks)

C-A* 2 The diagram below shows a pregnancy test stick.

Region containing monoclonal   Handle
antibodies to test for pregnancy hormone

Guided a) Describe what is meant by a monoclonal antibody.

Antibodies of one type produced in large quantities by...cells.

(1 mark)

b) Describe how monoclonal antibodies are produced.

....

....

....

....

(4 marks)

c) Suggest why this stick would not respond to a different hormone.

....

(1 mark)

The kidneys

E-C 1 The diagram below shows part of the urinary system.

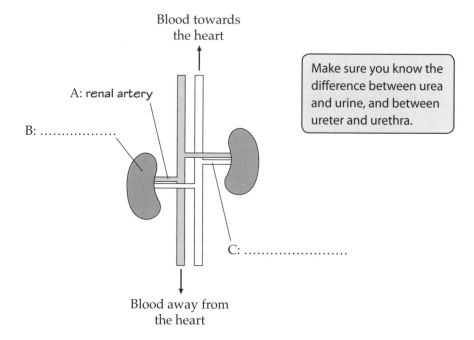

Blood towards
the heart

A: renal artery

B: ....

C: ....

Blood away from
the heart

Make sure you know the
difference between urea
and urine, and between
ureter and urethra.

Guided a) Fill in the missing labels on the diagram.

(3 marks)

b) Describe a function for structures B and C.

B: ....

C: ....

(2 marks)

c) Name the tubes that run from each kidney to the bladder.

....

(1 mark)

C-B 2 Describe how and where urea is produced.

Higher

....

....

....

(3 marks)

C-B 3 Sometimes a person's kidneys can fail. Describe the two possible treatments.

Higher Treatment 1: ....

....

Treatment 2: ....

....

(4 marks)

Inside the kidneys

  1 The nephron is involved in a range of processes. Match each box on the left to the correct box on the right using straight lines.

> Make sure you know the different parts of the nephron and what each part does.

Glomerulus and Bowman's capsule

Collecting duct

Filtration

Selective reabsorption of glucose

Reabsorption of water

(2 marks)

  2 The diagram below shows part of a nephron.

 Higher

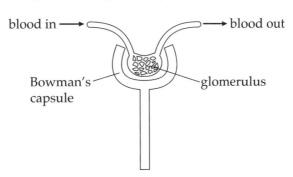

blood in ⟶   ⟶ blood out

Bowman's capsule

glomerulus

> Guided

a) Using the diagram, describe the movement of urea and a large protein within the glomerulus and Bowman's capsule.

Urea moves into the glomerulus and then ....

....

Large proteins do not pass into the .... because they

are to big to fit through the walls of the ....

(2 marks)

b) Suggest why red blood cells and glucose are not normally found in urine.

....

....

(2 marks)

  3 Suggest the importance of selective reabsorption in the nephron.

....

....

(2 marks)

Had a go ☐ **Nearly there** ☐ **Nailed it!** ☐

The role of ADH

Higher This whole page covers Higher material.

C–A* 1 A student took part in a 5 mile run on a hot day. He did not take any fluids with him to drink.

> **Guided**

 a) At the end of the run, the student needed to urinate. Describe how his urine would differ from that of a student who had drunk water during the same run.

The volume of urine would be ....

and the urine would be more...

(2 marks)

 b) Explain why the level of ADH changed in the student's blood during the run.

> This question is worth four marks which means that you are going to have to provide quite a lot of detail in your answer.

....

....

....

....

....

....

(4 marks)

C–A* 2 The hormone ADH is involved in osmoregulation.

 a) State three sites in the body where ADH would be found.

....

....

....

....

....

(3 marks)

 b) Explain why ADH is found at each site given in part **a)**.

....

....

....

....

....

(3 marks)

The menstrual cycle

E-D 1 a) The passage below is about part of the menstrual cycle. Complete the passage by filling in the missing word or words.

On average, the menstrual cycle lasts .... days. Day 1 is the first day of

.... After this the lining of the uterus starts to .... and on day

.... ovulation occurs. Ovulation is the release of an .... from the

....

(6 marks)

b) State the functions of FSH and LH.

....

(2 marks)

c) Name the other two hormones involved in the control of the menstrual cycle.

....

....

(2 marks)

D-A 2 The diagram below shows the timing of some features in a menstrual cycle.

a) Fill in the two missing labels on the diagram.

(2 marks)

b) Mark on the diagram the point at which fertilisation is most likely to occur.

....

(1 mark)

c) Use the diagram to give evidence that the person has become pregnant.

> The question asks you to use the diagram. It is a good idea to refer to the diagram in you answer.

....

....

(2 marks)

Had a go ☐ Nearly there ☐ Nailed it! ☐

Hormone control

Higher This whole page covers Higher material.

C-A*

Guided

1 The flow diagram below shows the sequence in which four hormones first increase in concentration in the body during the menstrual cycle.

| FSH | ⇨ | oestrogen | ⇨ | .... | ⇨ | .... |

a) Complete the flow diagram by adding in the missing hormones.

(3 marks)

b) State two functions of FSH in the menstrual cycle.

....

....

....

(2 marks)

c) Using the diagram and your own knowledge, suggest what would happen if oestrogen levels did not increase in the menstrual cycle.

....

....

....

....

(3 marks)

B-A*

2 Explain how the menstrual cycle is controlled by negative feedback.

> Negative feedback can also be seen in the control of water levels in the body by ADH.

....

....

....

....

....

....

(4 marks)

Fertilisation

E-D 1 Egg donation is one form of infertility treatment. What is meant by the term 'infertility'?

....

(1 mark)

D-B 2 **a)** The table below describes three infertility treatments. Complete the table by adding in the name of each infertility treatment.

> Guided

Description of infertility treatment	Name of infertility treatment
Joining of a sperm and an egg in a dish	
Chemicals are given to a woman to encourage ovaries to release eggs	
Name of woman into whom a genetically unrelated embryo is placed	surrogate

(4 marks)

b) Give one advantage and one disadvantage for the third infertility treatment given in the table.

....

....

(2 marks)

C-A 3 **a)** Sperm and eggs are sex cells. Compare the structure of a human sperm cell with a human egg cell. In your answer, give **two** similarities between these sex cells and **three** differences.

> Higher

....

....

....

....

....

(5 marks)

b) For two of the structural differences given in part **a)**, explain their function in fertilisation.

....

....

(2 marks)

C-A 4 Describe what happens when an infertile couple have a baby using egg donation.

> Higher

> This question asks you to describe something, which means you need to write down what happens – but you don't need to explain why each stage of the process is carried out.

....

....

....

....

(4 marks)

Sex determination

D-C 1 a) A baby girl is born. Explain which sex chromosome was in the sperm that fertilised the egg.

....

....

(2 marks)

⟩**Guided**⟩ b) i) Complete the Punnett square to show the sex chromosomes of both parents and all possible children.

> This is a Punnett square but you could also use a genetic diagram to show how X and Y chromosomes combine.

Father

	X	
Mother X		

(2 marks)

ii) State the phenotype of the child in the shaded box.

....

(1 mark)

D-B 2 a) A couple who have a girl wish to have a second child. Explain the chance of the couple's second child being a boy.

....

....

....

(3 marks)

b) Read this statement:

If a couple have had children and they are all girls, then the next child is more likely to be a boy.

Discuss whether you think this statement is correct.

....

....

....

(2 marks)

Sex-linked inheritance

Higher This whole page covers Higher material.

C-A*

1 Haemophilia is a sex-linked condition. Using haemophilia as an example, define sex-linked inheritance.

....

....

(2 marks)

C-A*

Guided

2 About 8% of men are red–green colour blind but only about 1% of women are red–green colour blind.

a) Using a Punnett square, explain how a woman could be colour blind.

> If you are commenting on a probability (for example the probability that a child will have red-green colour blindness) then you can use a percentage (25 %), a ratio (1:4) or a number (0.25).

	X^h	Y
X^h		
X^h		

	X^h	Y
X^H		
X^h		

....

....

....

....

(5 marks)

b) Suggest why it is more common for males to be colour blind than females.

....

....

....

(2 marks)

Biology extended writing 1

Photoperiodicity is important to plants. Using examples, explain why photoperiodicity is important to plants.

(6 marks)

> You will be more successful in extended writing questions if you plan your answer before you start writing.
>
> The question asks you to explain what photoperiodicity is and to give some examples.
>
> Think about:
> - what photoperiodicity is and where it is found
> - what effect photoperiodicity has
> - why these effects are important
> - using some examples to help with your answer.

....

....

....

....

....

....

....

....

....

....

....

....

....

....

....

....

....

....

....

....

....

Biology extended writing 2

Edward Jenner demonstrated that a little boy called James Phipps who had been previously infected with cowpox did not then catch smallpox. Explain how Edward Jenner's experiment prevented James Phipps from getting smallpox.

(6 marks)

> You will be more successful in extended writing questions if you plan your answer before you start writing.
>
> Check carefully the focus of this question. There are essentially two parts to the question. First describe what Jenner did to the boy and then explain the outcome.

....

....

....

....

....

....

....

....

....

....

....

....

....

....

....

....

....

....

....

....

....

....

....

Courtship

E-D
Guided

1 There are a number of different mating strategies shown by animals. Three of these are listed on the left. Match the mating strategies to the animals that use them by joining the boxes.

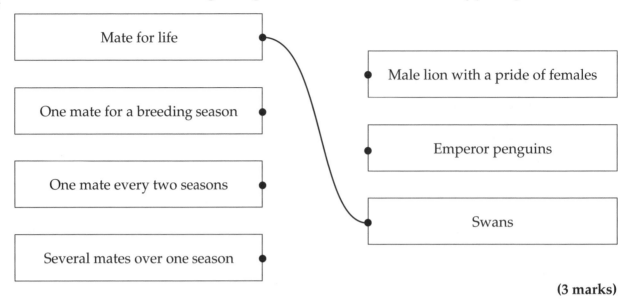

| Mate for life |
| One mate for a breeding season |
| One mate every two seasons |
| Several mates over one season |

| Male lion with a pride of females |
| Emperor penguins |
| Swans |

(3 marks)

D-C

EXAM ALERT

2 During courtship behaviour, usually one gender displays while the other selects. Complete the table for each of the examples given.

Example	Gender displaying	What is the selection being based on?
lion		
peacock		

(4 marks)

Students have struggled with exam questions similar to this
– **be prepared!**   ResultsPlus

In a recent question about courtship behaviours only 25% of students got full marks. Be clear about which sex is displaying and which is selecting. Usually (but not always) the female does the selecting.

C-A
Higher

3 In a number of bird species, the male birds dance for the female in a courtship display. She selects her mate based on the dance. Suggest how the dance helps the female to select a mate.

....

....

(2 marks)

Parenting

E-B　**1**　The saltwater crocodile is found in northern Australia. The female digs a hole by the edge of a river and lays her eggs in it. She covers the eggs with dead leaves. As the leaves decompose they release heat. The sex of the offspring is determined by the heat. If the nest is 30°C then all hatch out as females, and if the nest is at 32 °C then all of the offspring will be males.

　　a)　Suggest the ratio of newborn female to male saltwater crocodiles if the nest temperature was 31 °C.

　　..　**(1 mark)**

　　b)　Describe **one** way that the saltwater crocodile shows parental care.

> You do not need to know individual examples of parental care, but you do need to be able to apply what you know to new examples.

　　..

　　..　**(2 marks)**

D-A　**2**　Some poison dart frogs live in the Amazon rainforest. The female may lay her eggs on leaves rather than in water. The frogs then show a variety of parental care behaviours.

Guided　**a)**　Draw linking lines to match each description on the left to the correct parental care behaviour on the right. More than one description may link to the same behaviour.

(3 marks)

Description of poison dart frog parental care behaviour	Parental care behaviour example
Frog uses legs to spread water over eggs to keep them damp	Protecting the young from harm
After hatching, a parent carries the tadpoles to different places to separate them to stop the tadpoles eating each other	Helping the young to find food
	Sheltering the young from cold and wet
The female visits the tadpoles and lays unfertilised eggs for them to eat	Teaching the young new skills, such as hunting

Higher　**b)**　Suggest why these parental care behaviours can be a successful evolutionary strategy for poison dart frogs.

　　..

　　..　**(2 marks)**

C-A　**3**　Parental care is shown by many organisms in the animal kingdom. Some penguins in the
Higher　　Antarctic travel a long way from the sea to choose a mate, lay an egg and raise the chick.

　　a)　Describe why the parental care shown by these penguins can be a successful evolutionary strategy.

　　..

　　..　**(2 marks)**

　　b)　Suggest two risks for the parent penguins in trying to raise the chick.

　　..

　　..　**(2 marks)**

Simple behaviours

D-B
〉**Guided**〉

1   The table below describes four examples of animal behaviour. Complete the table by stating whether the behaviour is innate or imprinting.

Description of animal behaviour	Example is an innate behaviour or imprinting
Baby able to suckle for milk very soon after birth	innate
Young goose bonding with the first moving object it sees	
A chick using its 'tooth' on the end of its beak to crack open the shell so it can hatch out	
A butterfly being able to fly after emerging from its cocoon/chrysalis	

(4 marks)

b)   Explain what is meant by the term 'innate behaviour'.

....

(1 mark)

D-C

2   For each of the named people below, state the type of behaviour they studied.

Konrad Lorenz: ....   Niko Tinbergen: ....   **(2 marks)**

D-A*

3   Ten woodlice were placed in the centre of a choice chamber similar to the one shown.

After 10 minutes the number of woodlice in each half of the chamber was recorded. The study was repeated twice more and the results are presented in the table below.

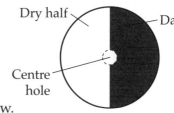

Dry half — Damp half
Centre hole

EXAM ALERT

a)   Complete the table by calculating the mean for each half.

Trial	Number of woodlice	
	Dry half	Damp half
1	0	10
2	1	9
3	0	10
Mean (average)		

Students have struggled with exam questions similar to this – **be prepared!**   ResultsPlus

Watch out! Sometimes two environmental factors are tested at the same time. In a recent question on this, less than 1 in 10 students identified both factors.

(1 mark)

Higher

b)   It was noticed that the woodlice moved more rapidly when in the dry half than when in the damp half. This form of movement is an innate behaviour. Explain how this innate behaviour led to the results for trial 1.

....

....

....

....

(4 marks)

Learned behaviour

E-B 1 a) Below is a list of four behaviours. Which two of the four behaviours can be explained by habituation? Place crosses in the boxes (☒) next to your answers.

☐ **A** getting used to a clock ticking so you do not hear it

☐ **B** suddenly running so you do not miss the bus

☐ **C** blinking your eye when a loud noise is heard

☐ **D** becoming used to wearing glasses   **(2 marks)**

b) Explain what is meant by the term 'habituation'.

....

....

....

(3 marks)

D-B 2 Experiments carried out on dogs by Pavlov showed classical conditioning.

Guided

EXAM ALERT

The statements on the left are four general statements, each describing one aspect of classical conditioning. On the right are four detailed descriptions of aspects of Pavlov's experiment. Link the general statements to descriptions using straight lines.

Innate response due to normal stimulus		Salivating when bell rung
Less salivation occurs as the animal becomes full		Bell means food presentation
An association of additional stimulus with normal stimulus		Salivating when presented with food
Conditioned behaviour shown		
Additional stimulus supplied		Bell rung with food presentation

(4 marks)

Students have struggled with exam questions similar to this – **be prepared!** ResultsPlus

Make sure you understand all the stages of classical conditioning.

C-A 3 The sea slug is an animal that lives in saltwater. It has gills on its surface for extracting oxygen from the seawater. If the gills are touched they withdraw into the animal. If this is repeated the gills eventually do not withdraw. This is an example of habituation. Suggest why this habituation is advantageous to the sea slug.

Higher

....

....

....

....

(4 marks)

Animal communication

E-D
Guided

1 Animals communicate using a number of different types of signal.

a) Give three different types of signal.

> You are not expected to learn specific examples of animal communication but it is a good idea to have an example that you can use.

Three different signal types include sound, ... and ....

(3 marks)

b) For one of the signal types given above, other than sound, give an example and state its function.

....

....

....

(2 marks)

C-A*
Higher

2 Jane Goodall and Dian Fossey both worked as ethologists.

> In this section of the specification you need to know about the work of different scientists. Make sure you know what each scientist worked on and what they discovered.

a) State the organism each ethologist worked with in their studies.

....

....

(2 marks)

b) Describe **three** ways in which their studies were similar.

....

....

....

(3 marks)

C-A*
Higher

3 Animals can communicate information to other animals using sound signals. Give examples of three different sound signals used by animals. For each one, explain what is being communicated.

....

....

....

....

....

(3 marks)

Plant communication

E-C   **1**   The diagram below shows a flower and four compass directions (north, south, east and west). The arrow shows the direction the wind was blowing from.

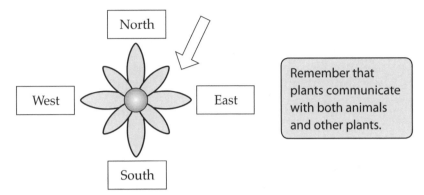

Remember that plants communicate with both animals and other plants.

 Guided

a) Assume that the plant communicates its presence by scent only. Which one of the four compass directions are most insects likely to come from? Explain your answer.

The insects are most likely to come from .... because

wind carries .... from that direction.

(2 marks)

b) Scent is an example of chemical communication in plants. Describe one other example of plants using chemicals for communication.

....

....

(2 marks)

c) Suggest two visual clues petals may have to communicate to pollinator insects.

....

....

(2 marks)

C-A*   **2**   Hummingbirds are highly adapted birds. They have long tongues to reach inside flowers

Higher   to feed. They can hover by beating their wings very fast, but this requires a lot of energy. They need even more energy during their breeding season. Some plants bloom during the hummingbird breeding season. The plants have long thin flowers with very sugar-rich nectar at the bottom. The structures that contain the pollen are carefully situated on the flowers.

Using examples from the passage above, explain why it is believed that the hummingbird and these plants have co-evolved.

....

....

....

....

(3 marks)

Human evolution

E-D

Guided

1   Using the diagrams of Ardi and Lucy, describe three differences between them.

The differences include:

Ardi   Lucy

   i)   they have different heights

   ii) ....

       ....

   iii) ....

       ....   **(3 marks)**

C-A*

Higher

2   Some evidence for human evolution has come from the fossil record of the brain. The table below shows some of this evidence.

> You do not need to remember details such as brain sizes but you do need to remember the names and the general trends.

Name of species	Year before present when species first appeared (millions of years ago)	Brain volume (cm^3)
Ardipithecus ramidus (Ardi)	4.4	350
Australopithecus afarensis (Lucy)	3.2	400
Homo habilis	2.4	550
Homo erectus	1.8	850

  a)   Describe the relationship between when each species first appeared and their brain volume.

      ....

      ....   **(2 marks)**

  b)   Suggest how sand could be used to help work out the brain volume of a fossil.

      ....

      ....   **(2 marks)**

  c)   The first stone tools are dated from about 2.4 million years ago. Using the table, suggest what may have enabled the use of stone tools.

      ....

      ....   **(2 marks)**

C-A*

Higher

3   The diagram shows two images of stone tools.

  a)   Suggest which one of these two stone tools is the oldest. Give reasons for your answer.

A   B

      ....

      ....

      ....   **(3 marks)**

  b)   Using the diagram, suggest how stone tool A was held. Give reasons for your answer.

      ....

      ....

      ....   **(3 marks)**

Human migration

E-D
Guided

1 Describe how climate change has helped humans to migrate out of Africa.

When the climate cooled down in the ice age, sea levels .... this meant that

there was less water and crossing out of Africa was ....

(2 marks)

C-A*
Higher

2 Mitochondrial DNA can be used as a tool to help investigate human migration and evolution.

a) State where mitochondrial are found within human cells.

....

(2 marks)

b) Mitochondrial DNA is more useful than nuclear DNA when investigating evolution. State two reasons why it is more useful.

1. ....

2. ....

(2 marks)

c) Explain why mitochondrial DNA is inherited down the female line.

> This question asks you to explain something so you need to say *what* is happening and *why*.

....

....

....

(3 marks)

d) Explain how mitochondrial DNA supports the African Eve theory.

....

....

....

....

(3 marks)

e) The gene for having hairy ears is carried on the Y chromosome. What is the likelihood of this gene being seen in the mitochondrial DNA of the descendants of African Eve? Explain your answer.

....

....

....

(3 marks)

Biology extended writing 3

Using examples, explain how previous climate changes have affected human behaviour.

(6 marks)

> You will be more successful in extended writing questions if you plan your answer before you start writing.
>
> In this case, make sure you follow the requirements of the question. Here the question is asking you to:
>
> - link climate change to human behaviour
> - give examples to illustrate your answer.

....

....

....

....

....

....

....

....

....

....

....

....

....

....

....

....

....

....

....

....

....

....

....

....

....

Biotechnology

E-C

1 The diagram shows a fermenter used to make a substance called penicillin.

Guided

a) Fill in the three missing labels on the diagram. **(3 marks)**

b) Explain the function of the water jacket.

....

....

....

....

(2 marks)

c) Explain the function of the sterile air.

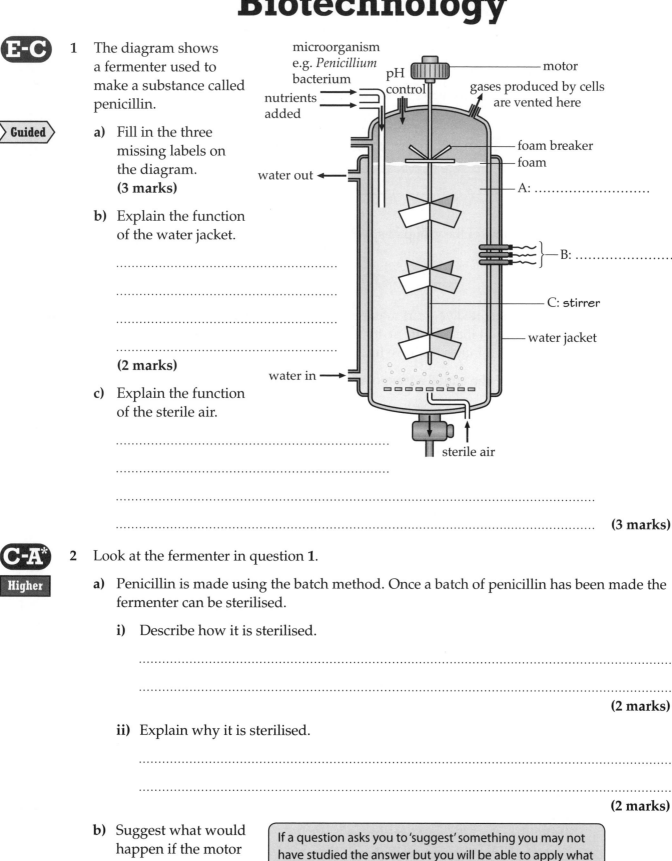

microorganism e.g. *Penicillium* bacterium

nutrients added

pH control

motor

gases produced by cells are vented here

foam breaker

foam

water out ◄—

A: ....

B: ....

C: stirrer

water jacket

water in ——►

sterile air

....

....

....

.... **(3 marks)**

C-A*

Higher

2 Look at the fermenter in question **1**.

a) Penicillin is made using the batch method. Once a batch of penicillin has been made the fermenter can be sterilised.

   i) Describe how it is sterilised.

   ....

   .... **(2 marks)**

   ii) Explain why it is sterilised.

   ....

   .... **(2 marks)**

b) Suggest what would happen if the motor stopped working.

> If a question asks you to 'suggest' something you may not have studied the answer but you will be able to apply what you know to a new situation and come up with an answer.

....

....

.... **(3 marks)**

Microorganisms for food

E–C

Guided

1 a) Complete the table below by writing the type of microorganism that makes each of the foods given in the left-hand column.

Name of food produced by microorganism	Type of microorganism
bread	
mycoprotein	fungus
wine	
yoghurt	

(4 marks)

b) Give the name of the group of species used to make mycoprotein.

....

(1 mark)

D–A

Higher

2 A student investigated yoghurt making. She added a teaspoon of yoghurt to some sterilised milk and then kept the mixture at 40 °C. She placed a pH probe in the mixture and recorded the pH every 30 minutes for three hours.

a) Explain why she added a teaspoon of yoghurt at the start.

....

(1 mark)

b) Suggest why she used sterilised milk.

....

....

....

(3 marks)

c) The table below shows the change in pH of the mixture over the three hour (180 minute) study. Yoghurt contains milk proteins that have thickened due to a fall in pH.

Time/min	0	30	60	90	120	150	180
pH	6.4	6.4	6.2	6.0	5.6	5.3	5.1

i) Explain the changes to the milk that the student would observe during the three hour study.

....

....

....

(3 marks)

ii) Explain the relationship between time and pH during yoghurt making.

Use the data in the table to help you.

....

....

....

....

(5 marks)

Mycoprotein

1 Mycoprotein can be made into a variety of foods, such as a burger. The table below shows nutritional information on mycoprotein and minced beef.

Food	Energy content (cal/100 g)	Fat (g/100 g)	Saturated fat (g/100 g)	Cholesterol (mg/100 g)
Mycoprotein (uncooked)	146	12	0	0.0
Minced beef (uncooked)	291	16	6	76.0

Guided

  a) Using the data for minced beef in the table, calculate the percentage of fat that is saturated fat. Show your working.

Percentage of saturated fat = amount of saturated fat ÷ total amount of fat × 100%

> Always show your working.

Percentage .... %

(2 marks)

  b) Explain what the minced beef cholesterol figure shows.

....

....

....

(3 marks)

  c) Explain one possible health issue that can arise if too much high energy food is eaten.

....

(1 mark)

2 Mycoprotein is made from a fungus grown in a fermenter.

  a) The fungus that makes mycoprotein is small. Give three advantages of using microorganisms for food production.

....

....

....

(3 marks)

  b) Suggest why ammonia is supplied to the fungus in the fermenter.

....

....

(2 marks)

  c) Suggest why air is supplied to the fungus in the fermenter.

....

....

(2 marks)

Had a go ☐ **Nearly there** ☐ **Nailed it!** ☐

Enzyme technology

D–A* 1 Explain why immobilising an enzyme may be beneficial for the manufacturer.

....

....

(2 marks)

C–B 2 Give the reaction which is catalysed by the enzyme invertase.

Guided Invertase catalyses the breakdown of .... into glucose

Higher and ....

(2 marks)

C–A*

Higher 3 Chymosin is an enzyme used in enzyme technology.

a) Describe how chymosin is made.

....

....

....

(3 marks)

b) Describe the function of chymosin.

....

....

(2 marks)

c) Natural chymosin was obtained from calves' stomachs. Suggest one benefit of using chymosin produced by genetically modified microorganisms.

....

(1 mark)

> You need to know the names and uses of the enzymes named in the specification.
> They are **chymosin** and **invertase**.

4 Why are enzymes added to low temperature washing powders?

....

....

(2 marks)

DNA technology

Higher This whole page covers Higher material.

1 The flow diagram below shows some stages in making human insulin.

Stage 1: Restriction enzyme used on human DNA	Stage 2: Restriction enzyme used to open plasmid from a bacterium	Stage 3: DNA ligase used to form the modified plasmid	Stage 4: Modified plasmid placed into bacteria and bacteria placed in a fermenter

a) State why the bacteria are placed in a fermenter in stage 4.

....

(1 mark)

Guided **b)** Describe the function of the restriction enzyme in stage 1 of this process.

The restriction of enzymes cut out the...

for insulin and leave...which means the

gene can splice together with the bacterial...

(3 marks)

c) Explain why the same restriction enzyme should be used in stage 2.

....

....

(2 marks)

d) Explain what is meant by the term 'modified plasmid' in stage 3.

....

....

(2 marks)

e) Suggest why it is sensible to try to have only bacteria with modified plasmids in the fermenter.

....

....

....

(3 marks)

Exam questions similar to this have proved especially tricky in the past – **be prepared!** ResultsPlus

Make sure you are clear what happens in each step of this process. In a recent question on this, only one quarter of students got full marks. The plasmid must be removed from a bacterium before the gene is inserted. The plasmid is then inserted into another bacterium.

Global food security

D-B

Guided

1 a) The table below shows two different pest management strategies. Complete the table by explaining how each works.

Pest management strategy	How it works
Using pheromone traps	Pheromone attracts pests to the trap
Attracting natural predators	

(2 marks)

Higher

b) Give **two** reasons why it is better to use a combination of pest management strategies rather than just using a pesticide.

....

....

(2 marks)

> A **biofuel** is a fuel that is produced from living organisms, such as oil palm trees.

D-B

2 a) State **two** advantages of replacing fossil fuels with biofuels.

....

....

(2 marks)

b) Explain a disadvantage of replacing fossil fuels with biofuels.

....

....

(2 marks)

B-A*

3 Food production can be increased by conventional plant breeding programmes.

Higher

a) Suggest three different characteristics that could be selected for in a crop suitable for use in any country.

....

....

....

(3 marks)

b) Suggest two other characteristics that might be selected for in a crop to be grown in a hot, dry part of Africa.

....

....

(2 marks)

c) Suggest why wheat in the United Kingdom has been selected to have a short stem length.

....

....

(1 mark)

A GM future?

C–A

Higher

1 a) What does the term 'transgenic plant' mean?

....

....

....

(2 marks)

b) What is meant by the term 'vector'?

....

....

....

....

(3 marks)

D–A*

Guided

2 Tomato plants can be genetically modified to contain substances called flavonoids.

a) Suggest why the production of tomato fruits containing flavonoids may be of interest to people with cancer.

Some evidence from experiments show that mice with cancer live longer if given

.... (therefore people with cancer) may want to eat these tomatoes

because the flavonoids may ....

(2 marks)

b) State why the production of tomato fruits containing flavonoids may not benefit all people with cancer.

....

(1 mark)

Higher

c) *Agrobacterium tumefaciens* can be genetically modified to contain a recombinant plasmid with the gene for making flavonoids. Describe how these bacteria can be used to produce a tomato plant that will produce tomato fruits rich in flavonoids.

> You need to learn the process of making a GM plant.

....

....

....

....

....

....

(5 marks)

Had a go ☐ Nearly there ☐ Nailed it! ☐

Insect-resistant plants

Higher This whole page covers Higher material.

C–A* 1 Bt plants are an example of insect-resistant plants. Bt plants have had a gene from the *Bacillus thuringiensis* bacterium introduced into them.

Guided a) Describe the function of Bt toxin.

The Bt toxin is a chemical that is poisonous ....

....

....

(3 marks)

b) Describe how bacteria are used to produce plants that produce the Bt toxin.

....

....

....

....

(3 marks)

C–A* 2 There are advantages and disadvantages to using transgenic plants such as the insect-resistant plant containing the Bt gene.

a) Explain **two** advantages.

1. ....

2. ....

(2 marks)

b) One disadvantage may be the incorporation of the Bt gene into wild plants.

i) Explain how this could occur.

....

....

....

(2 marks)

ii) Suggest why this would be considered a disadvantage.

....

....

....

(2 marks)

> You need to understand that new scientific developments have advantages, but also disadvantages.

Biology extended writing 4

Explain the health and nutritional advantages of using mycoprotein as a food source.

(6 marks)

> You will be more successful in extended writing questions if you plan your answer before you start writing.
>
> In this question you are being asked to:
> - describe the nutritional advantages of using mycoprotein as a food source
> - explain the health advantages of mycoprotein.

....

....

....

....

....

....

....

....

....

....

....

....

....

....

....

....

....

....

....

....

....

....

....

Had a go ☐ Nearly there ☐ Nailed it! ☐

Biology extended writing 5

Explain how the genetic material of a bacterium is changed through recombinant DNA technology.

(6 marks)

You will be more successful in extended writing questions if you plan your answer before you start writing.

In this example, the process you are being asked to describe has a definite sequence of events, so it would be sensible for you to write your response in this order. Make sure you have used correct technical words, such as 'restriction enzymes' and 'ligase'.

....

....

....

....

....

....

....

....

....

....

....

....

....

....

....

....

....

....

....

....

....

....

....

....

Water testing

 E-D
Guided

1   Flame tests are often used to identify the ions present in a substance. Draw lines to link the ion present to the colour of the flame it would produce.

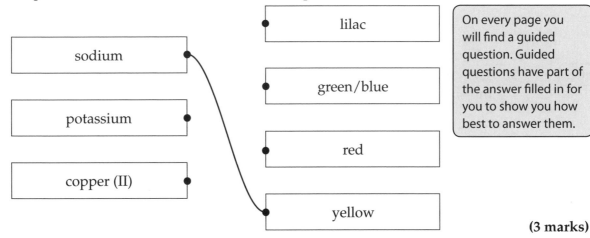

| sodium |
| potassium |
| copper (II) |

| lilac |
| green/blue |
| red |
| yellow |

On every page you will find a guided question. Guided questions have part of the answer filled in for you to show you how best to answer them.

(3 marks)

D-B
EXAM ALERT

2   Scientists can use flame tests or precipitation tests as qualitative tests on unknown substances.

Students have struggled with exam questions similar to this in the past – **be prepared!** ResultsPlus

The words 'qualitative' and 'quantitative' are very similar – you need to know the difference between the two.

a)   What is meant by the term **qualitative**?

....

(1 mark)

b)   The calcium ion can be identified by a flame test. State the colour that calcium ions give to a Bunsen flame.

....

(1 mark)

c)   Calcium ions can also be identified using sodium hydroxide solution. Describe what you would see when sodium hydroxide solution is added to a solution containing calcium ions.

....

....

(2 marks)

 B-A*
Higher

3   Iron can form two different chlorides: iron (II) chloride ($FeCl_2$) and iron (III) chloride ($FeCl_3$). Maggie has a solution of each chloride. She adds sodium hydroxide solution to both solutions.

a)   State the observation she will make in each case:

   i)   iron (II) chloride ....

   ii)   iron (III) chloride...

(2 marks)

b)   Write a balanced chemical equation for the reaction between iron (II) chloride solution and sodium hydroxide solution. Give the state symbols.

....

(3 marks)

35

Safe water

D-B 1 A student is testing a solution to see if it contains bromide ions.

a) Name the solution he needs to use to test for bromide ions.

....

(1 mark)

⟩**Guided**⟩ **b)** The student needs to add another substance before he adds the solution in **a)**. State what this substance is and why it is needed.

He needs to add .... acid. This acid reacts with any

.... ions present, as these ions also give a precipitate with silver nitrate.

(2 marks)

c) State what he would see in his test if bromide ions were present.

....

(1 mark)

D-B 2 Many fertilisers are ammonium compounds.

a) Give the formula of the ammonium ion...

(1 mark)

b) Name the solution used in the test for the ammonium ion.

> The test for the ammonium ion is more complex than most other ion tests. This question breaks the test up into steps. In an exam, you may have to give the test with no prompts.

....

(1 mark)

c) When it is warmed with the solution in **b)**, the ammonium ion gives off a gas. Name this gas and describe how you would test for it.

Name of gas ....

Test...

....

(2 marks)

B-A* 3 Drinking water contains chloride ions. Most of these ions come from the chemicals used to treat drinking water. Silver nitrate is used to test for chloride ions in the water.

EXAM ALERT

> Students have struggled with exam questions similar to this
> – **be prepared!**   ResultsPlus

a) Name the precipitate formed when silver nitrate reacts with chloride ions.

....

(1 mark)

b) Write a balanced ionic equation for the reaction producing this precipitate.

....

(2 marks)

Safe limits

D-C 1 Use words from the box to complete the sentences about testing water purity.

dissolved	lilac	molten	yellow
precipitate	red	solution	

Scientists test water to find out if it has .... ions in it. Some tests are flame

tests – for example, potassium ions turn a flame .... Other tests involve

adding sodium hydroxide solution and looking for a coloured ....

(3 marks)

C-A
Higher

2 The drug lithium carbonate is used to treat people with depression. Doctors need to monitor the level of the drug present in the blood of a patient.

a) Lithium is in the same group of the periodic table as sodium and potassium. The same test is used for lithium ions as for sodium ions and potassium ions. Give the name of this test.

....

(1 mark)

Guided

b) A known volume of blood is tested for the carbonate ion. Hydrochloric acid is added to the blood and a gas is produced.

 i) Describe how you would identify this gas.

 I would bubble the gas through a solution of ....

 The solution would turn ....

(2 marks)

 ii) The volume of gas produced is measured. Explain whether this test is qualitative or quantitative.

 ....

 ....

(2 marks)

B-A*
Higher
EXAM ALERT

3 Two chemists are testing some water. They add two drops of sodium hydroxide solution to a sample of the water. A white precipitate forms. One chemist says that calcium ions are present; the other says aluminium ions are present.

> Students have struggled with exam questions similar to this in the past – **be prepared!** ResultsPlus

a) Explain why this confusion might occur.

....

(1 mark)

b) Describe how a further test could show who is correct.

....

....

(2 marks)

Water solutes

G-E 1 Some dissolved substances make water hard.

 Guided Complete the table by putting a tick (✓) or a cross (✗) to show whether the ion listed makes water hard.

Ion	calcium	sodium	chloride	magnesium	sulfate
Does it make water hard?	✓				

(4 marks)

D-B 2 The water in a student's town is hard. When she uses soap, she notices a precipitate form in the water.

a) Name this precipitate. ....

(1 mark)

b) Explain why this precipitate forms.

....

....

(2 marks)

EXAM ALERT **c)** Describe another observation that she would make when using soap.

> Students have struggled with exam questions similar to this in the past – **be prepared!** **ResultsPlus**

> Remember to say what you actually see.

....

....

(2 marks)

d) The student finds that 2 dm³ of the water in her town contains 5 g of calcium sulfate. What is the concentration of the calcium sulfate in g dm⁻³?

Concentration .... g dm⁻³
(2 marks)

B-A* 3 A student has 250 cm³ of a solution of sodium chloride that is labelled as 25 g dm⁻³.

 Higher **a)** Calculate the mass of sodium chloride dissolved in this solution.

Mass ....
(2 marks)

b) The student adds 750 cm³ of pure water to the solution of sodium chloride. Calculate the concentration of the diluted solution that the student has made.

Concentration .... g dm⁻³
(2 marks)

Hard and soft water

G-E 1 Water with dissolved calcium and magnesium compounds is called hard water.

   a) One type of hardness is called temporary hardness.

      i) State how temporary hardness can be removed from water.

....

                                                              (1 mark)

      ii) When temporary hardness is removed, a white solid is formed. Give the name of this solid.

....

                                                                (1 mark)

   b) The other type of hardness is called permanent hardness. Name a substance that causes permanent hardness in water.

....

                                                                (1 mark)

D-B 2 Many houses have a water softener attached to their water supply to remove hard water. Many of these water softeners contain ion exchange resins.

   a) Explain one reason why people might want to soften their water.

....

....

                                                             (2 marks)

Guided **b)** Describe how an ion exchange resin softens hard water.

     The resin contains .... ions, held on ....

     which swap places with...ions.

                                                                (3 marks)

   c) After some time, the ion exchange resin stops working effectively. Suggest how the resin could be made effective again.

....

                                                                (1 mark)

B-A* 3 Explain how boiling softens temporary hard water. Include a balanced equation in your answer.

Higher

> Your equation needs to show the correct chemical formulae. Some formulae that will be useful to you are $CaCO_3$ and $Ca(HCO_3)_2$.

....

....

....

....

                                                                (4 marks)

Moles and mass

Higher This whole page covers Higher material.

Relative atomic masses: H = 1, C = 12, N = 14, O = 16, Mg = 24, K = 39, Ca = 40

C-B **1** **a)** Work out the relative formula mass of water. Place a cross in the box (☒) next to your answer.

☐ **A** 8 ☐ **B** 18 ☐ **C** 32 ☐ **D** 34 **(1 mark)**

b) Which of the following contains 1 mole of particles? Place a cross in the correct box (☒) next to your answer.

☐ **A** 8 g of oxygen gas ☐ **B** 12 g of carbon

☐ **C** 20 g of magnesium ☐ **D** 40 g of potassium **(1 mark)**

C-B **2** This question is about carbon dioxide (CO_2).

Guided

a) Calculate the relative formula mass of carbon dioxide.

Relative formula mass = (1 × ......) + (2 × ......) = ...

(1 mark)

b) How many moles of carbon dioxide are there in the following?

Remember, if you use the formula shown in the guidance below, the mass needs to be **in grams**.

i) 88 g of carbon dioxide

number of moles = mass ÷ relative formula mass

= 88 g ÷ ....

Amount .... moles
(1 mark)

ii) 88 kg of carbon dioxide

Amount .... moles
(2 marks)

B-A* **3** You have 84 g of calcium carbonate ($CaCO_3$) and 84 g of calcium oxide (CaO). Explain which sample has the larger number of moles of substance.

....

....

....

(3 marks)

B-A* **4** Hydrogen gas forms molecules (H_2). A balloon contains 2 g of hydrogen gas. Explain the difference between the numbers of moles of hydrogen gas and the number of moles of hydrogen atoms in the balloon.

....

....

....

(3 marks)

Moles in solution

 This whole page covers Higher material.

 1 A student is making up several solutions. Each solution must have a concentration of $1\,mol\,dm^{-3}$. Complete the table to show the mass of each solute that she needs to be able to make up $1\,dm^3$ of solution. (Relative atomic masses: $H = 1$, $O = 16$, $Na = 23$, $S = 32$, $Cl = 35.5$)

Solute	Mass needed (in g)
sodium hydroxide (NaOH)	
hydrogen chloride (HCl)	
sulfur dioxide (SO$_2$)	

(4 marks)

 2 Limewater is a saturated solution of calcium hydroxide (Ca(OH)$_2$) in water. $1\,dm^3$ of limewater contains $1.85\,g$ of calcium hydroxide. (Relative atomic masses: $H = 1$, $O = 16$, $Ca = 40$)

a) What is the concentration of the limewater in $g\,dm^{-3}$?

$$\text{Concentration} = \frac{\text{mass dissolved}}{\text{volume in dm}^3}$$

Concentration .... $g\,dm^{-3}$

(1 mark)

b) Calculate the number of moles in $1.85\,g$ of calcium hydroxide.

relative formula mass of Ca(OH)$_2$ = ... + (2 ×......) + (2 ×......) = ...

number of moles = mass ÷ relative formula mass

= ....g ÷ ....

Number of moles ....

(2 marks)

c) Therefore, state the concentration of limewater in $mol\,dm^{-3}$.

Concentration .... $mol\,dm^{-3}$

(1 mark)

> You will find this formula useful:
> concentration (in $mol\,dm^{-3}$) = moles of substance ÷ volume (in dm^3)
> Remember that $1\,dm^3 = 1000\,cm^3$, so to convert cm^3 to dm^3 you must divide by 1000.

3 a) Calculate which of these solutions of sodium hydroxide (NaOH) has the greatest concentration in $mol\,dm^{-3}$: $20\,g$ of sodium hydroxide in $500\,cm^3$ of water, or $32\,g$ of sodium hydroxide in $750\,cm^3$ of water.

Answer ....

(4 marks)

b) Calculate which of these solutions of hydrogen chloride (HCl) has the greatest number of moles of HCl: $200\,cm^3$ of a $0.5\,mol\,dm^{-3}$ solution, or $500\,cm^3$ of a $0.25\,mol\,dm^{-3}$ solution.

Answer ....

(3 marks)

Preparing soluble salts 1

G-E 1 Draw lines to link the name of the acid to the salts that it produces.

| sulfuric |
| nitric |
| hydrochloric |

| chlorate |
| nitrate |
| chloride |
| sulfate |

(3 marks)

D-B 2 Some students are making a salt. They add nickel carbonate to sulfuric acid a bit at a time. The nickel carbonate reacts with the acid, and fizzes. Eventually, the fizzing stops and some solid nickel carbonate remains in the bottom of the beaker.

a) Name the gas given off in the reaction. ....
(1 mark)

Guided **b)** Explain why the fizzing stops.

As the reaction goes on, all the ....

is used up, so the .... stops.
(2 marks)

c) State how the unreacted nickel carbonate can be removed from the solution.

....
(1 mark)

d) Write a word equation for the reaction.

....
(2 marks)

B-A* 3 A student has some silver oxide (Ag_2O) and wants to make some silver salts. The student adds excess silver oxide to a warm solution of nitric acid, then removes any unreacted silver oxide.

Higher

a) Write a balanced chemical reaction for the reaction to make this salt.

....
(3 marks)

> Your first step needs to be to work out the name of the salt formed here, and its formula.

b) Describe how the student can make crystals of this salt.

....

....
(2 marks)

c) Suggest why the student cannot use this method to make silver chloride.

....
(1 mark)

Preparing soluble salts 2

G-E
Guided

1 The diagram shows the apparatus used to react nitric acid with sodium hydroxide solution. Complete the labels to name the equipment used.

(4 marks)

....

nitric acid

sodium hydroxide + indicator

D-B

2 Potassium chloride can be made by reacting potassium hydroxide solution (KOH) with hydrochloric acid (HCl). A student measured out 25.0 cm³ of potassium hydroxide solution, put it into a conical flask, and neutralised it with hydrochloric acid.

a) What piece of apparatus is used to measure 25.0 cm³ of potassium hydroxide?

....

(1 mark)

b) Describe how this piece of apparatus is used.

....

....

....

(2 marks)

> Students have struggled with exam questions similar to this in the past – **be prepared!** ResultsPlus

> It is important that you can say what the pieces of apparatus used in a titration are, and why each of these pieces is useful.

c) Write a balanced chemical equation for the reaction.

....

(2 marks)

B-A*
Higher

3 A student is making sodium chloride. She reacts 25.0 cm³ of sodium hydroxide solution with hydrochloric acid. She finds that it takes 24.0 cm³ of hydrochloric acid to neutralise the sodium hydroxide solution. When she repeats the experiment, it takes 26.4 cm³ of hydrochloric acid.

a) Explain how she will know the reaction is over.

> To answer this question, you need to say what you are going to use, and how you would use it to show that the neutralisation reaction is over.

....

....

(2 marks)

b) Write an ionic equation to show what is meant by the word **neutralisation**.

....

(1 mark)

c) Explain how she should work out the volume of hydrochloric acid needed to neutralise 25.0 cm³ of sodium hydroxide solution.

....

....

....

(3 marks)

Titration calculations

Higher This whole page covers Higher material.

 C–B

Guided

1 Sodium nitrate can be prepared by reacting together solutions of sodium hydroxide and nitric acid. A student did a titration to find the volume of nitric acid needed to react with 25.0 cm³ of sodium hydroxide solution.

   a) Complete chemical equation for this reaction.

   NaOH + ... → .... + H_2O **(2 marks)**

   b) The diagram shows the readings on the burette at the start and at the end of a titration. Use these diagrams to complete the table.

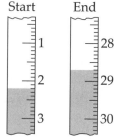

Start   End

Reading at end	28.7 cm³
Reading at start	cm³
Volume of solution added	cm³

   (3 marks)

B–A* **2** In a titration reaction, 25.0 cm³ of lithium hydroxide solution is reacted with hydrochloric acid.

   LiOH (aq) + HCl (aq) + → LiCl (aq) + H_2O (l)

A student recorded concordant titration results for the hydrochloric acid needed: 26.50 cm³ and 26.70 cm³

   a) Calculate the average volume of acid added.

   ....

   (1 mark)

> 'Concordant' is a chemistry word for results that are very close together. In titrations, results are concordant if they are within 0.20 cm³ of each other.

   b) Use the concordant results to calculate the average volume of acid added.

   ....

   (1 mark)

   c) In another reaction, 25.0 cm³ of 0.50 mol dm⁻³ lithium hydroxide solution was neutralised by 27.50 cm³ of hydrochloric acid.

EXAM ALERT

> Students have struggled with exam questions similar to this in the past – **be prepared!** **ResultsPlus**

> This calculation is structured for you to get you into the habit of setting out your answer and showing all your working. In any calculations, you can score marks for your working even if your final answer is wrong.

   i) Calculate the amount of lithium hydroxide solution used in moles.

   Amount .... moles

   (2 marks)

   ii) Calculate the amount, in moles, of hydrochloric acid used.

   Amount .... moles

   (1 mark)

   iii) Calculate the concentration, in mol dm⁻³, of the hydrochloric acid.

   Amount .... moles

   (2 marks)

More calculations from equations

Higher This whole page covers Higher material.

 1 Calculate the number of moles of substance in each of the following solutions.

Guided

    a) 25.0 cm³ of 0.40 mol dm⁻³ calcium hydroxide solution

$$\text{number of moles} = \frac{\text{volume in cm}^3}{1000} \times \text{concentration}$$

                                           Amount .... moles

                                                     (1 mark)

    b) 20.0 cm³ of 0.50 mol dm⁻³ copper sulfate solution

                                           Amount .... moles

                                                     (1 mark)

    c) 50.0 cm³ of 0.15 mol dm⁻³ sulfuric acid

                                           Amount .... moles

                                                     (1 mark)

 2 A titration can be used to find out the concentration of calcium ions in a solution of hard water. A sample of hard water is reacted with sodium carbonate solution. The calcium ions in the hard water react with the carbonate ions to make a solid precipitate of calcium carbonate.

$$Ca^{2+}(aq) + CO_3^{2-}(aq) \rightarrow CaCO_3(s)$$

25.0 cm³ of hard water reacts with 21.5 cm³ of 0.015 mol dm⁻³ sodium carbonate solution.

    a) Calculate the amount, in moles, of sodium carbonate used.

                                           Amount .... moles

                                                    (2 marks)

    b) State the number of moles of calcium ions in the hard water. Explain how you got your answer.

        ....

        ....

                                                              (2 marks)

    c) Calculate the concentration of calcium ions in hard water, in mol dm⁻³.

                                   Concentration .... mol dm⁻³

                                                      (2 marks)

B-A* **3** In a titration, 25.0 cm³ of 0.200 mol dm⁻³ potassium hydroxide solution was neutralised by 28.40 cm³ of sulfuric acid. The equation for the reaction is:

$$2KOH + H_2SO_4 \rightarrow K_2SO_4 + 2H_2O$$

Calculate the concentration, in mol dm⁻³, of the sulfuric acid.

> If you get stuck, look at the steps in question **2** and follow the same pattern. Remember to show all your working.

                                          Concentration .... mol dm⁻³

                                                     (3 marks)

Chemistry extended writing 1

Aluminium sulfate, $Al_2(SO_4)_3$, is used during the treatment of water. It is important to check for the presence of this substance in the water leaving the treatment plant to make sure that the level in drinking water is safe.

Describe how you would test qualitatively and quantitatively for aluminium sulfate in drinking water. You should include balanced equations for any tests that you describe.

(6 marks)

You will be more successful in extended writing questions if you plan your answer before you start writing.

In this question, you are asked to test for a single compound. To plan your answer you should:

- split the compound into its constituent ions – and give a test for each one
- remember to give the substances used in the test and a result for each test
- remember what makes the test for the aluminium ion different from tests for other metal ions – some tests for metal ions give very similar results
- make sure that you know the difference between a qualitative and a quantitative test, and clearly show which part of your answer belongs to which test.

Remember that quantitative tests ask you to say how much of the substance is there – you'll need to think about how to use all your chemistry knowledge to suggest an answer here.

....

....

....

....

....

....

....

....

....

....

....

....

....

....

....

....

....

....

....

Chemistry extended writing 2

Neutralisation reactions are used to produce metal salts.

Explain why different methods are used to make the soluble salts copper (II) chloride and sodium chloride through neutralisation reactions.

(6 marks)

You will be more successful in extended writing questions if you plan your answer before you start writing.

The introduction to this question contains the information you need to help you answer. Remember that neutralisation reactions involve an acid and a base. You need to contrast the two different methods that you have studied to make salts.

For each method, you need to:

- give the starting materials to make the salt
- relate the solubility of the metal base chosen to the method chosen
- give enough detail of the methods to show that there are differences
- include any other substances that will be needed in the reaction
- show how you can tell that the reaction is complete.

Note that this question is not a 'compare' question – so there is no need to give long descriptions of the methods. You are asked instead to explain the difference – so you only need enough information to show how the methods and different and say why they are different.

....

....

....

....

....

....

....

....

....

....

....

....

....

....

....

....

....

....

Electrolysis

D-B   **1**   A student places a crystal of purple potassium manganate (VII), $KMnO_4$, on a piece of filter paper on a glass slide. He connects a d.c. electricity supply to each end of the paper.

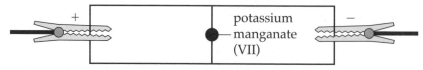

> Most potassium compounds are white or colourless, but potassium manganate (VII) is purple because it contains the coloured manganate ion.

When he turns on the power supply, nothing happens. He turns the power supply off, wets the filter paper, and switches the power supply on again. After some time, he notices a purple streak moving to the left.

a) Explain why nothing happens when the paper is dry, but something does happen when the paper is wet.

....

....

(2 marks)

> **Guided**   **b)** Explain what causes the purple streak.

The streak is caused by the .... ion, which is .... charged, so

moves towards the .... electrode.

(2 marks)

B-A*   **2**   A student is electrolysing a solution of hydrochloric acid. She collects samples of gases at both electrodes during the electrolysis.

Higher

a) Hydrogen forms at the negative electrode, according to the equation $2H^+ + 2e \rightarrow H_2$. State whether this is an oxidation or a reduction reaction. Give a reason for your answer.

....

(1 mark)

b) Write an equation for the formation of the product at the positive electrode.

....

....

(2 marks)

c) Explain how you would expect the volumes of the gases produced at the electrodes to compare to each other.

....

....

....

(3 marks)

Making and using sodium

1 Sodium is produced by the electrolysis of molten sodium chloride in a special electrolysis cell, at a temperature of 600°C. The chlorine is produced at the positive electrode. It is kept separate from the sodium produced at the other electrode.

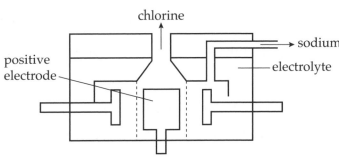

a) Complete the equation showing the formation of chlorine in the electrolysis.

$2Cl^- \rightarrow$ .... + ...

(1 mark)

b) Explain why the chlorine produced is kept separate from the sodium.

> Some questions may ask you to explain details that you have learned, even though you may not have learned a reason at the time. However, you should be able to use your chemical knowledge to work out the answer. They're not difficult questions, just a bit unfamiliar!

If the chlorine and the sodium come together they would ....

This would make ....

(2 marks)

c) Sodium metal melts at 98°C and boils at 883°C. Explain the physical state of the sodium produced during the electrolysis.

....

....

(2 marks)

d) Give a use for the sodium produced.

....

(1 mark)

2 Like sodium, potassium is also made by electrolysis. In one factory, molten potassium bromide is used as the electrolyte.

a) State what is meant by the term **electrolyte**.

....

(1 mark)

b) Potassium is in the same group of the periodic table as sodium. Write an equation showing the formation of potassium at the negative electrode.

....

(2 marks)

c) When the potassium is removed from the electrolysis cell, it is kept under an atmosphere of argon until it cools down. Suggest why this is done.

....

....

(2 marks)

49

Electrolysis of salt water

 1 Use words from the box to complete the sentences about the electrolysis of salt water.

| anions | brine | cations | chloride | hydrogen | limewater |

Salt water is often called .... It is a solution of sodium chloride. When

it is electrolysed, negative .... ions travel to the positive electrode to

make chlorine gas. Hydrogen gas is made at the other electrode as the positive hydrogen

ions, known as ...., are discharged.

(3 marks)

D-B 2 Complete the table to show the products made when different solutions are electrolysed.

> Guided >

Some of the solutions here may be unfamiliar to you, but you can apply what
you know about the solutions that you have studied to these unfamiliar ones.

Solution	Product at cathode	Product at anode
copper (II) chloride		chlorine
copper (II) nitrate	copper	
potassium bromide	hydrogen	
sodium sulfate		oxygen

(4 marks)

 3 Sodium chloride solution can be electrolysed to form a variety of products. The electrolysis
cell contains a mesh to separate the two electrodes from each other.

a) State what you would see at the
positive electrode during electrolysis.

....

....

This question is NOT asking you to
name the product formed – it is asking
you what observation you would make
as the electrolysis takes place.

(1 mark)

b) Although the solution contains positive sodium ions, the product at the negative
electrode is hydrogen. Explain why this product is formed.

....

....

(2 marks)

c) Write an equation for the formation of hydrogen at the negative electrode.

....

(2 marks)

d) Some litmus paper is dipped in the solution around the negative electrode. It turns blue.
Explain this observation.

....

....

(2 marks)

More electrolysis

1 Complete the labels on the cell used in the purification of copper.

pure copper cathode, electrolyte, cell, anode, sludge   Use theses words to help you.

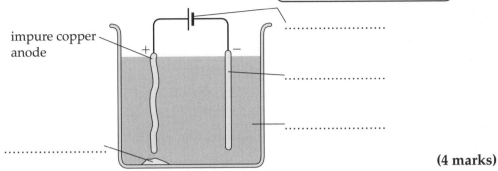

impure copper
anode

....

....

....

....

(4 marks)

2 Components on a circuit board are often made from copper, but the copper is then electroplated with gold.

   a) What would you choose for the two electrodes in this electroplating?

     Positive electrode:...

     Negative electrode:...

(2 marks)

   b) The electroplating solution contains gold ions, Au^{3+}.

     i) Write an equation for the conversion of these gold ions into gold atoms.

     ....

(1 mark)

     ii) Explain why this process is described as a reduction.

     ....

(1 mark)

B-A*
Higher

3 A student electrolyses a solution of copper (II) sulfate using copper electrodes. He measures the mass of the negative electrode before and after the experiment. He records his results in a table.

| Mass of electrode before experiment | 10.25 g |
| Mass of electrode after experiment | 10.63 g |

EXAM ALERT

   a) Explain why the negative electrode increases in mass, illustrating your answer with an equation.

Students have struggled with exam questions similar to this in the past – **be prepared!** ResultsPlus

....

....

Take care with your answer here. Be very clear about using words like 'atom' and 'ion' so that the examiner can see that you know exactly what is going on in this reaction.

....

....

(3 marks)

   b) The student measured the mass of the positive electrode before the experiment. Its mass was 9.78 g. Estimate the mass of this electrode after the experiment, explaining your reasoning.

....

....

(2 marks)

Gas calculations 1

Higher This whole page covers Higher material.

Note: in all these questions the gases are at room temperature and pressure.

C-B   **1**   Complete the table to show the volumes and numbers of moles present in samples of gases. Some have been done for you.

> Remember that there are 1000 cm^3 in 1 dm^3.

Gas	Volume	Number of moles
carbon dioxide		0.5
ethane	1200 cm^3	

(2 marks)

C-B   **2**   A student has two empty balloons. She puts 0.5 moles of helium gas into one balloon.

   a)   What volume of helium is in the balloon?

Volume .... dm^3

(1 mark)

Guided   **b)**   She then puts 4800 cm^3 of hydrogen into the second balloon. How many moles of hydrogen gas are in this balloon?

I mole of gas occupies 24 dm^3, or 24 000 cm^3, so 2400 cm^3 is 0.1 moles, and

4800 cm^3 is .... moles.   **(1 mark)**

B-A*   **3**   One mole of a substance contains 6×10^{23} particles. A student has 1 mole of chlorine gas (Cl_2).

   a)   How many chlorine molecules are there in the sample?

....

(1 mark)

   b)   How many moles of chlorine atoms are there in the sample? Explain your answer.

....

....

(2 marks)

B-A*   **4**   One step in the manufacture of sulfuric acid is the reaction between sulfur dioxide and oxygen to make sulfur trioxide.

$$2SO_2\,(g) + O_2\,(g) \rightarrow 2SO_3\,(g)$$

   a)   A factory starts with 960 dm^3 of sulfur dioxide gas. Calculate how many moles of gas are in this volume.

Answer .... moles

(1 mark)

   b)   Explain what the minimum number of moles of oxygen would be needed to react with this sulfur dioxide.

....

....

(2 marks)

   c)   In a different reaction, 100 dm^3 of sulfur dioxide gas is reacted with 50 dm^3 of oxygen gas. Explain the volume of sulfur trioxide that would be made.

....

....

(2 marks)

Gas calculations 2

 Higher This whole page covers Higher material.

C-A

Guided

1 In a blast furnace, iron (III) oxide is reduced by carbon monoxide gas to produce metallic iron.

$$Fe_2O_3(s) + 3CO(g) \rightarrow 2Fe(l) + 3CO_2(g)$$

1600 g of iron (III) oxide are placed into a blast furnace.

a) How many moles of iron (III) oxide are present in 1600 g?
(Relative atomic masses: O = 16, Fe = 56)

Formula mass of iron (III) oxide = (2 × 56) + (3 × 16) = ...

Number of moles of iron (III) oxide = $\dfrac{\text{mass in grams}}{\text{formula mass}}$ = ...

= .... moles

(2 marks)

b) How many moles of carbon dioxide gas will this produce?

> The numbers in front of the chemical formulae tell you the ratio in which the substances react in the reaction. Many candidates lose marks by not using these numbers from the chemical equation.

One mole of iron (III) oxide produces three moles of carbon dioxide

So, moles of carbon dioxide = 3 × moles of iron (III) oxide

= .... moles

(1 mark)

c) Calculate the volume of carbon dioxide produced.

Volume .... dm³

(2 marks)

B-A*

2 A student is planning to investigate the reaction of magnesium with hydrochloric acid and wants to measure the volume of hydrogen produced in this reaction, using a 100 cm³ gas syringe.

$$Mg(s) + 2HCl(aq) \rightarrow MgCl_2(aq) + H_2(g)$$

Calculate the maximum mass of magnesium ribbon
the student can use, if she wants to make no more
than 100 cm³ of gas.

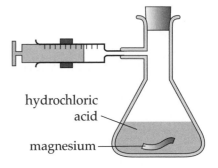

hydrochloric
acid

magnesium

Mass .... g

(4 marks)

Fertilisers

G-E   1   Use words from the box to complete the sentences about fertilisers.

ammonia	animals	manure	nitrogen
oxygen	paper	plants	polymers

Fertilisers are substances that help .... grow. Some natural fertilisers

exist, such as .... Artificial fertilisers can also be made. The most

important element they contain is .... Many artificial fertilisers are

made from ....

(4 marks)

D-B   2   Fritz Haber invented the process by which ammonia is made industrially. The process
involves reacting hydrogen and nitrogen together.

a)   State the raw materials used to provide the hydrogen and nitrogen.

....

....

(2 marks)

> **Guided**   b)   Give a balanced chemical equation for the reaction.

......$N_2(g)$ + ...$H_2(g)$ ⇌ ......$NH_3(g)$

(2 marks)

c)   Explain what is shown by the symbol ⇌ in the equation.

....

....

(2 marks)

C-A*   3   One common fertiliser is ammonium nitrate, NH_4NO_3.

Higher   a)   Write a balanced equation for the formation of ammonium nitrate from ammonia and
nitric acid.

....

(2 marks)

b)   Explain why fertilisers are useful.

....

....

(2 marks)

c)   Ammonium nitrate, like all nitrates, is soluble in water. Explain why this can cause
problems if farmers spread too much fertiliser on their fields.

....

....

....

....

(4 marks)

Equilibrium

Higher This whole page covers Higher material.

C-B   1   The table shows some equilibrium reactions. Place a tick beside the equation that has the fewer gas molecules.

$N_2(g) + 3H_2(g)$		$\rightleftharpoons$	$2NH_3(g)$	
$2SO_2(g) + O_2(g)$		$\rightleftharpoons$	$2SO_3(g)$	
$4NH_3(g) + 5O_2(g)$		$\rightleftharpoons$	$4NO(g) + 6H_2O(g)$	

(3 marks)

C-A   2   The reaction between hydrogen and iodine to form hydrogen iodide is shown by the equation:

$$H_2(g) + I_2(g) \rightleftharpoons 2HI(g)$$

This reaction forms a dynamic equilibrium.

a)   Explain what is meant by the term **dynamic equilibrium**.

....

....

> Remember to explain both words in this term: **dynamic** and **equilibrium**.

(2 marks)

 b)   The reaction is endothermic in the forward direction. What effect would an increase in temperature have on the equilibrium yield of hydrogen iodide? Explain your answer.

An increase in temperature favours the ....thermic reaction.

Here, this is the ....ward reaction, so the yield is ....

(2 marks)

c)   Explain why changing pressure has no effect on the position of equilibrium for the reaction.

....

....

(2 marks)

B-A*   3   Nitrogen dioxide, NO_2, is a brown gas. The gas molecules react with each other to form dinitrogen tetroxide, N_2O_4, which is colourless. The reaction is a reversible reaction.

$$2NO_2(g) \rightleftharpoons N_2O_4(g)$$

a)   Explain what effect increasing pressure would have on this equilibrium.

....

....

(2 marks)

b)   An equilibrium mixture of the two gases is placed in a sealed tube. The tube is placed in very cold water. The colour of the gases gets paler. Explain whether the forward reaction is exothermic or endothermic.

....

....

....

(3 marks)

The Haber process

Higher This whole page covers Higher material.

C-B
Guided

1 Ammonia is made using the Haber process. Describe the conditions used in this reaction.

The temperature used is .... °C

The pressure used is .... atm

and the catalyst used is ....

(3 marks)

C-A

2 One step in the manufacture of sulfuric acid is the production of sulfur trioxide:

$$2SO_2(g) + O_2(g) \rightleftharpoons 2SO_3(g)$$

The reaction uses vanadium(V) oxide as a catalyst.

a) What effect does the catalyst have on the rate of this reaction?

....

(1 mark)

b) Explain the effect that the catalyst has on the position of equilibrium.

....

....

(2 marks)

B-A*

3 The graph shows how the yield of ammonia made in the Haber process depends on the temperature and pressure used in the reaction.

a) Using the graph, state the ideal conditions for making the maximum yield of ammonia.

....

....

....

....

(2 marks)

b) The pressure and temperature that you identified in **a)** are not used in the reaction.
Explain why not.

> To answer this question you need to give reasons – in terms of rate and cost – why the conditions used are not those suggested in **a)**; and how the real conditions vary.

....

....

....

....

(4 marks)

c) Explain why a catalyst is used in this reaction.

....

....

(2 marks)

Chemistry extended writing 3

Copper is used in electrical circuit boards and in wiring, as it is a very good conductor of electricity. However, impurities in copper reduce its conductivity, so it is very important to be able to manufacture copper in a very pure state. This purification of copper is done using an electrolytic process.

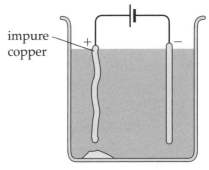
impure copper

Explain how this electrolytic process, involving redox reactions, leads to the formation of pure copper. You should give balanced equations to illustrate your answer.

(6 marks)

> You will be more successful in extended writing questions if you plan your answer before you start writing.
>
> There are a few clues in the diagram – hopefully, with only one part of the cell labelled, you can identify other key parts of the cell. In your answer, you should:
>
> - use clear, correct terminology, especially when describing the electrodes
> - explain what happens at each of the two electrodes
> - write half-equations to represent the reactions at the electrodes
> - use the half-equations to describe the reactions as oxidation or reduction
> - give some information about why the copper produced is pure – probably the simplest way to do this is to say what happens to the impurities.
>
> Remember 'OILRIG' as you come to think about the redox processes here –
>
> Oxidation Is the Loss of electrons; Reduction Is the Gain of electrons.

....

....

....

....

....

....

....

....

....

....

....

....

....

....

....

Fermentation and alcohol

G-E **1** Answer the following questions about fermentation. Put a cross (☒) in the box next to your answer.

    a) To make ethanol, a solution of glucose is fermented using:

      ☐ **A** carbon dioxide   ☐ **B** starch   ☐ **C** water   ☐ **D** yeast

          (1 mark)

    b) As well as ethanol, fermentation produces:

      ☐ **A** carbon dioxide   ☐ **B** starch   ☐ **C** water   ☐ **D** yeast

          (1 mark)

    c) What is the optimum temperature for fermentation?

      ☐ **A** 20°C   ☐ **B** 40°C   ☐ **C** 60°C   ☐ **D** 80°C

          (1 mark)

D-B **2** Beer is made by fermenting the sugar present in barley. The fermenting beer is kept in anaerobic conditions and the temperature is carefully controlled.

    a) What is meant by the term **anaerobic**?

    ....

          (1 mark)

    b) Explain why it is important to control the temperature of the fermentation mixture.

    ....

    ....

          (2 marks)

B-A*

Higher

Guided

3 The following drinks all contain about two units of alcohol.

> One way to answer this question is to use the percentage data to work out the number of millilitres of alcohol in each drink, and then compare the two.

Pint of beer   Glass of wine   Double vodka
550 ml   150 ml   50 ml
3.5% alcohol   12.5% alcohol   37.5% alcohol

    a) Show, by calculation, that the amount of alcohol in the pint of beer is almost the same as the amount of alcohol in the glass of wine.

    Beer: 550 ml and 3.5% alcohol, so volume of alcohol = $550 \times \dfrac{3.5}{100}$ = ...

    Wine: 150 ml and 12.5% alcohol ....

    ....

          (3 marks)

    b) Wine is made by fermentation. Write the chemical equation for fermentation.

        $C_6H_{12}O_6 \rightarrow$ .... C_2H_5OH + ...   **(2 marks)**

    c) Describe how drinks with higher alcohol concentrations, such as vodka, are made.

    ....

    ....

          (2 marks)

Ethanol production

Higher This whole page covers Higher material.

C–B

Guided

1 Complete the table showing the names and formulae below.

Structure	Name	Formula
$H-O-H$		
$\begin{array}{c} H \qquad\quad H \\ \diagdown \qquad \diagup \\ C=C \\ \diagup \qquad \diagdown \\ H \qquad\quad H \end{array}$		C_2H_4
$\begin{array}{c} H \quad H \\ \vert \quad\; \vert \\ H-C-C-O-H \\ \vert \quad\; \vert \\ H \quad H \end{array}$	ethanol	

(6 marks)

C–A

2 Ethanol can be made by this reaction: $C_2H_4(g) + H_2O(g) \rightarrow C_2H_5OH(g)$

   a) Name the raw material used to provide the ethene for this reaction.

   ....
   (1 mark)

   b) Describe how the ethene is obtained from this raw material.

   > There are two processes for you to consider here. You don't need to give many details, but you should name the processes and say very briefly what each does.

   ....

   ....

   ....

   ....
   (4 marks)

   c) Ethanol can be converted back into ethene. What type of reaction does this involve?

   ....
   (1 mark)

   d) What do the state symbols in the reaction tell you about the temperature used in the process? Explain your answer.

   ....

   ....
   (2 marks)

B–A*

3 Zambia is a country in southern Africa. The soil is quite fertile and coffee is grown in Zambia. However, the country has only small reserves of crude oil. A factory is opening that needs ethanol to use as a solvent to decaffeinate the coffee. Explain which process of making ethanol is likely to be of more use in Zambia.

   ....

   ....

   ....

   ....
   (4 marks)

Homologous series

1 Draw lines to link the formula of the compound with its name. All the compounds are members of the alkane or alkene homologous series.

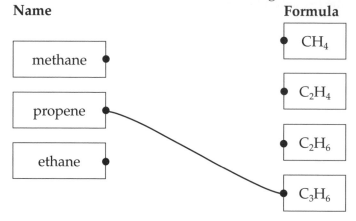

Name   Formula

(3 marks)

2 The compound butane, C_4H_{10}, is a member of the homologous series of alkanes.

a) Draw the structure of a butane molecule.

> Students have struggled with exam questions similar to this in the past – **be prepared!** ResultsPlus

> Remember to count up the hydrogens – and to show **all** the carbon–hydrogen bonds.

(2 marks)

b) Describe what is meant by the term **homologous series**.

....

....

....

(3 marks)

3 The first three members of the homologous series of alcohols are methanol (CH_3OH), ethanol (C_2H_5OH) and propanol (C_3H_7OH).

a) Use this information to suggest a general formula for the alcohol homologous series.

....

(2 marks)

b) Ethanol reacts with phosphorus(V) chloride to make a compound with the formula C_2H_5Cl. Predict the formula of the compound made when phosphorus(V) chloride reacts with propanol.

....

(1 mark)

c) The complete combustion of ethanol produces carbon dioxide and water. Use this information to write a balanced equation for the complete combustion of methanol.

....

(2 marks)

Ethanoic acid

G–E

Guided

EXAM ALERT

1 Complete the following word equations for the reactions of ethanoic acid.

Students have struggled with exam questions similar to this in the past – **be prepared!** ResultsPlus	Remember that ethanoic acid reacts in the same way as any other acid, so think about the products made when acids react with metals and with carbonates.

     a) magnesium + ethanoic acid → magnesium ethanoate + ...

     b) zinc carbonate + ethanoic acid → zinc .... + ... + water

                                                                   (4 marks)

D–B

2 A student opens a bottle of wine. He leaves the wine in a warm part of the kitchen. When he comes back to the bottle after several days, he notices that the wine tastes sour. The sour taste is due to ethanoic acid.

     a) Draw the structure of ethanoic acid.

                                                                   (2 marks)

     b) Explain how this ethanoic acid has been made in the wine.

       ....

       ....

                                                                   (2 marks)

B–A*

Higher

3 The first three members of the homologous series of carboxylic acids are methanoic acid ($HCOOH$), ethanoic acid (CH_3COOH) and propanoic acid (C_2H_5COOH).

     a) Use this information to suggest a formula for the next carboxylic acid in the series.

       ....

                                                                   (1 mark)

     b) The boiling points of ethanoic acid and propanoic acid are 118°C and 141°C. Predict the boiling point of butanoic acid, the next member of the homologous series. Explain how you got your answer.

       ....

       ....

                                                                   (2 marks)

     c) A student adds some sodium carbonate, Na_2CO_3, to a solution of propanoic acid. One product of this reaction is sodium propanoate (C_2H_5COONa). Write a balanced equation for the reaction.

       ....

       ....

                                                                   (2 marks)

Esters

G-E 1 Answer the following questions about polyesters. Put a cross (☒) in the box next to your answer.

a) One use of polyesters is to make plastics for:
☐ **A** drink bottles ☐ **B** food wrap ☐ **C** paper ☐ **D** ropes

b) In order to conserve raw materials, waste polyesters are:
☐ **A** buried ☐ **B** burned ☐ **C** decomposed ☐ **D** recycled

c) Polyesters that are re-used to make fabrics are often turned into:
☐ **A** fleece jackets ☐ **B** milk crates ☐ **C** ropes ☐ **D** silk

D-B 2 Ethyl methanoate is an ester made from ethanol and methanoic acid. It has the smell of rum.

Guided Describe how this ester can be used commercially.

This ester can be used as a .... in the .... industry because of its characteristic taste and smell.

(2 marks)

B-A* 3 Propanoic acid and ethanol react together to form an ester.
$$C_2H_5COOH + C_2H_5OH \rightleftharpoons C_2H_5COOC_2H_5 + H_2O$$

Higher

a) Name the ester that is made in this reaction.

....

(1 mark)

b) Draw the structural formula of this ester, showing all bonds.

> Remember that a structural formula should show all bonds – this includes all the C–H bonds in this molecule. Be very careful showing what happens at the ester link.

(2 marks)

c) The reaction which makes this ester is slightly exothermic. Suggest what effect a very high temperature has on the yield of the ester.

....

....

(2 marks)

d) Some people describe esterification as being similar to the neutralisation of an acid, such as nitric acid, to make a salt. Give one similarity and one difference between esterification and the making of a salt.

....

....

(2 marks)

Fats, oils and soap

G-E 1 Use suitable words to complete the sentences about fats and oils.

Oils are liquids at room temperature, and fats are .... Both oils and fats

are .... These molecules can be broken down by heating with a strong

solution of .... to make ....

(4 marks)

D-B 2 Margarine is made from vegetable oil. To make margarine, the vegetable oil is heated with hydrogen.

a) What else is needed to make this reaction take place?

....

(1 mark)

Guided b) Describe the chemical reaction that takes place when hydrogen reacts with the vegetable oil.

The reaction is called .... which means that the

C=C bonds break and .... atoms are added.

(2 marks)

c) State what happens to the melting point of the oil as it turns into margarine.

....

(1 mark)

B-A* 3 a) The diagram shows a typical molecule found in soap. Use it to explain the meanings of the terms **hydrophobic** and **hydrophilic**.

Higher

....

....

(2 marks)

b) Soaps are made from oils or fats. Describe how soaps are made from these substances.

....

....

(2 marks)

c) Explain how soaps remove greasy stains from our clothes.

EXAM ALERT

Students have struggled with exam questions similar to this in the past – **be prepared!** **ResultsPlus**

Be very careful with the terms that you use – part **a)** should have helped you think about these. You need to think about what each part of the soap molecule does.

....

....

....

(3 marks)

Chemistry extended writing 4

Many millions of tonnes of ammonia are made in industry each year. This ammonia can be used to make some plastics, as well as nitric acid and dyes. However, most of the ammonia is used to make artificial fertilisers.

Evaluate the impact of fertilisers on people and the environment.

(6 marks)

> You will be more successful in extended writing questions if you plan your answer before you start writing.
>
> The question here asks you to 'evaluate' the impact of fertilisers. To answer this type of question, you need to:
> - identify what fertilisers do that is positive
> - explain why these positives are necessary
> - identify some negative aspects of fertiliser use
> - explain why these negative aspects arise.
>
> Finally, and most importantly for an 'evaluate' question, you should attempt to balance the positive and negative aspects that you have examined to describe whether using fertilisers has an overall positive or negative effect.

....

....

....

....

....

....

....

....

....

....

....

....

....

....

....

....

....

....

....

....

Chemistry extended writing 5

The diagram shows a typical soap molecule.

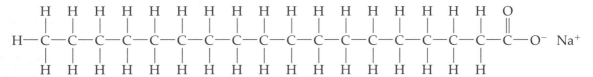

Explain how the structure of soap relates to the substances it is made from, and the way in which it works in removing grease.

(6 marks)

> You will be more successful in extended writing questions if you plan your answer before you start writing.
>
> This question has some straightforward parts, but some aspects are more difficult. Don't spend too long on describing how soap is made – most candidates will get some marks here. Make your answer of a higher level by addressing the more difficult aspect of how soap works.
>
> Your answer should include:
>
> - the substances that soap is made from
> - a brief description of how this works – try to concentrate on what happens *chemically*, not just on describing how you would do the experiment
> - a description of the key features of the soap molecule
> - how these features allow the soap to interact with grease on our clothes or skin.
>
> Don't forget to use good scientific language. Words that you should try to include are: ester, hydrolysis, carboxylic acid, hydrophilic, hydrophobic.

....

....

....

....

....

....

....

....

....

....

....

....

....

....

....

....

Radiation in medicine

F-D

Guided

1   Medical physicists help doctors to check patients using a variety of techniques that involve radiation. Draw a line linking the type of radiation to its use.

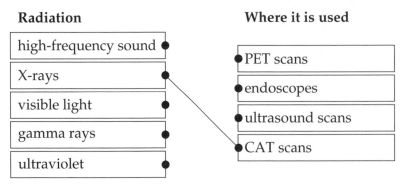

Radiation

| high-frequency sound |
| X-rays |
| visible light |
| gamma rays |
| ultraviolet |

Where it is used

| PET scans |
| endoscopes |
| ultrasound scans |
| CAT scans |

On every page you will find a guided question. Guided questions have part of the answer filled in for you to show you how best to answer them.

(4 marks)

D-B

2 a) Explain why the following are types of radiation.

Think about what they carry and where they come from.

| visible light   alpha radiation   ultrasound   X-rays |

....

....

(2 marks)

b) i)   State which type of radiation is the odd one out.

....

(1 mark)

ii) Give a reason for your answer.

....

(1 mark)

c) Which of the radiations shown in the box is ionising?

....

(2 marks)

D-B

3   Explain why people who stand close to the loud speakers at rock concerts are more likely to damage their hearing than those who stand several metres away.

....

....

(2 marks)

B-A*

Higher

4   On a sunny day a solar panel of area 1.6 m² receives solar power of 2000 W. Calculate the intensity of the sunlight. State the unit of your answer.

$$\text{Intensity} = \frac{\text{power of incident radiation}}{\text{area}}$$

The formula will be given at the beginning of the exam paper.

Intensity .... unit ....

(3 marks)

How eyes work

F-D 1

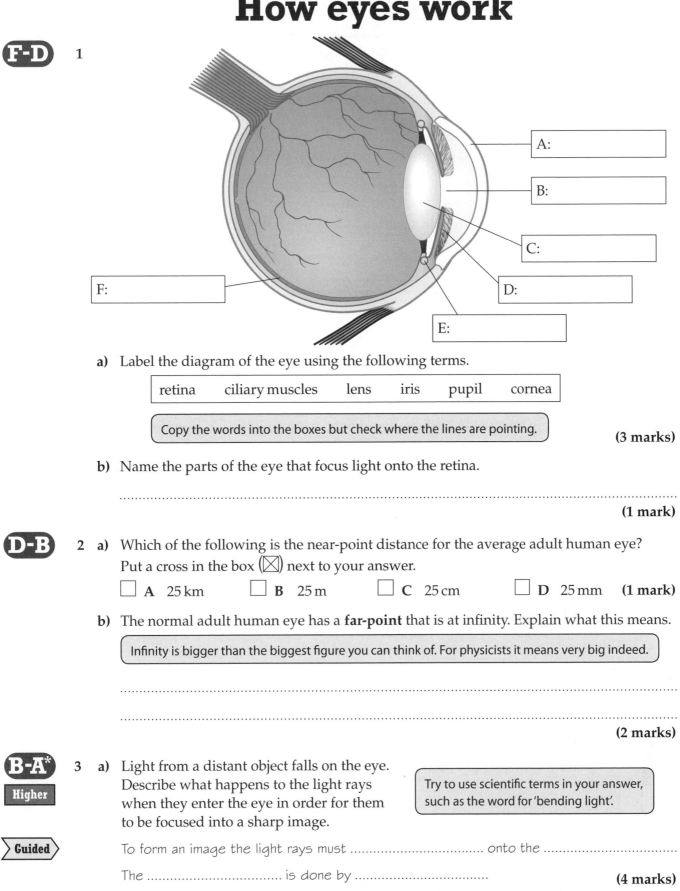

A:

B:

C:

D:

E:

F:

a) Label the diagram of the eye using the following terms.

| retina   ciliary muscles   lens   iris   pupil   cornea |

> Copy the words into the boxes but check where the lines are pointing.

(3 marks)

b) Name the parts of the eye that focus light onto the retina.

....

(1 mark)

D-B 2 a) Which of the following is the near-point distance for the average adult human eye?
Put a cross in the box (☒) next to your answer.

☐ **A** 25 km   ☐ **B** 25 m   ☐ **C** 25 cm   ☐ **D** 25 mm   **(1 mark)**

b) The normal adult human eye has a **far-point** that is at infinity. Explain what this means.

> Infinity is bigger than the biggest figure you can think of. For physicists it means very big indeed.

....

....

(2 marks)

B-A*

Higher

3 a) Light from a distant object falls on the eye.
Describe what happens to the light rays
when they enter the eye in order for them
to be focused into a sharp image.

> Try to use scientific terms in your answer,
> such as the word for 'bending light'.

> **Guided**

To form an image the light rays must .... onto the ....

The .... is done by ....   **(4 marks)**

b) Describe what happens in the eye to allow it to focus on an object that is close.

....

....

(2 marks)

Sight problems

F-D 1 Underline one word in each pair of words in the following sentences to make each sentence a true statement.

A person with *long/short* sight has blurred vision of objects that are near because their eyeballs are too *long/short*, so that the image is formed *inside/outside* the eye instead of on the retina. A person with long sight may also have eye lenses that cannot get *thin/fat* enough to form a sharp image. A person with short sight may have a cornea that is curved too *little/much*.

(5 marks)

D-B 2 The diagram shows a lens in spectacles correcting a sight problem and producing a sharp image on the retina.

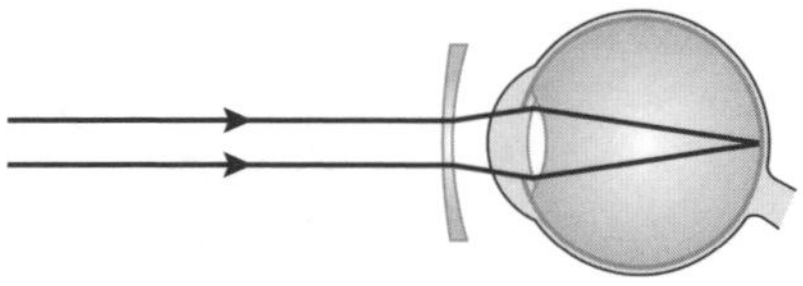

a) State the sight problem being corrected in the diagram.

....

(1 mark)

Guided b) Explain why the sight correction shown in the diagram is successful.

The lens makes the light rays .... so that the image moves

....

(2 marks)

c) i) State a precaution that must be taken if using contact lenses instead of spectacles to correct sight problems.

....

(1 mark)

ii) Contact lenses allow oxygen to pass through them. Give a reason why this is necessary.

| Remember that the cornea is made up of living cells. |

....

....

(1 mark)

B-A* 3 a) Describe how lasers can be used to correct some types of sight problem.

Higher

| You do not need to give full details of the surgery needed. |

....

....

(2 marks)

b) Suggest one advantage and one disadvantage of laser treatment compared with other methods of sight correction.

....

....

(2 marks)

Correcting sight problems

Many people have either long sight or short sight. Explain the causes of these problems and how they can be treated.

You may include diagrams in your answer.

> Some questions to think about are:
> - What happens to light when it enters the eye?
> - What makes someone short or long sighted?
> - What methods are available to correct long and short sight?
> - How do these methods work?

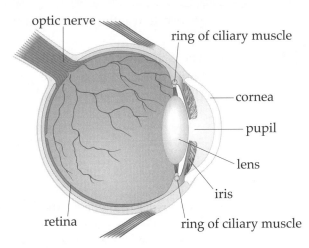

optic nerve
ring of ciliary muscle
cornea
pupil
lens
iris
ring of ciliary muscle
retina

(6 marks)

> This question will also test your quality of written communication. That means you need to write a clear, balanced and detailed answer.

Guided When light enters the eye it is bent by the lens and the cornea. This means that an image forms on the retina. In short sight ....

....

....

....

....

....

....

....

....

....

....

....

....

....

....

....

....

....

Different lenses

You may need to use the formula: power of a lens $= \dfrac{1}{\text{focal length}}$

   1 a) The diagram shows parallel rays of light
being refracted by a converging lens.
Label the diagram to show the focal point.

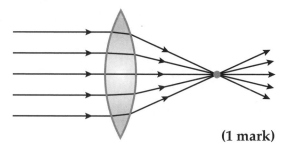

(1 mark)

b) Use a ruler to measure the focal length of the lens in the diagram.

> Mark on the diagram where you
> are making the measurement.

Focal length ....

(1 mark)

c) A different lens has a shorter focal length than the one shown in the diagram. State how
the shape of this lens would differ from the one in the diagram.

....

(1 mark)

   2 A converging lens can be used to focus sunlight to set fire to a piece of paper. Explain why
this cannot be done with a diverging lens.

> **Guided**

When rays of sunlight pass through a diverging lens they ....

so the rays ....

(2 marks)

   3 a) A student has two lenses. Lens A has a focal length of 0.50 m and lens B has a focal
length of 0.25 m. Which of the following statements is correct? Put a cross in the
box (☒) next to your answer.

☐ **A** Lens A has the highest power because it has the flattest faces.

☐ **B** Lens B has the highest power because it has the flattest faces.

☐ **C** Lens A has the highest power because it has the more sharply curved faces.

☐ **D** Lens B has the highest power because it has the more sharply curved faces.

(1 mark)

b) Calculate the power of lens B.

> When calculating power of a lens
> the focal length must be in metres.

Power ....dioptres

(2 marks)

   4 A lens in a pair of spectacles has a power of 3.75D. Calculate the focal length of the lens.

> **Higher**

Focal length ....m

(2 marks)

The lens equation

Higher This whole page covers Higher material.

You will need this formula in these questions: $\dfrac{1}{f} = \dfrac{1}{u} + \dfrac{1}{v}$

1 A student carried out an experiment to find the focal length of a converging lens. She placed a light bulb 100 cm from the lens. A sharp image was formed when the student held a screen 25 cm from the lens on the opposite side from the lamp.

   a) Explain why the image formed is a real image.

....

....

(2 marks)

   b) Calculate the focal length of the lens.

> All the distances must have the same unit. You can use centimetres in this question.

....

....

Answer ....

(3 marks)

Guided

2 The focal length of an adult human eye is 22 mm. The distance from the lens to the retina is 25 mm. Calculate the distance from an object to the eye's lens that will produce a sharp image on the retina.

$f = 22\,\text{mm},\ v = 25\,\text{mm}$

$\dfrac{1}{u} = \dfrac{1}{f} - \dfrac{1}{v} =$

Answer ....

(3 marks)

3 The lenses in a pair of spectacles have a focal length of −40 cm. An object is 100 cm from the spectacles.

> Take care with + and − signs in the calculation.

   a) Calculate the distance from the spectacles to the image.

....

....

....

(3 marks)

   b) State the type of image formed by the lens.

....

(1 mark)

Reflection and refraction

 1 The diagram shows a ray of light hitting a surface and being reflected off it. Draw and label the normal, label the angle of incidence and the angle of reflection.

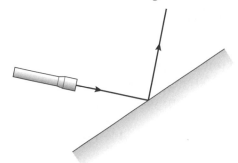

(3 marks)

 2 A ray of light hits a surface at an angle of 50° to the surface. Calculate the angle of reflection.

> Remember the number of degrees in a right angle.

Angle ....

(2 marks)

 3 A child sees a coin in a pool of water, but when she reaches in to pick it up it is not where she expected it to be. Explain why the child was mistaken.

> Guided

A ray travelling from the coin to the surface of the pool was bent..

from the normal because the ray was ....

(2 marks)

 4 Gamma rays enter a block of aluminium at an angle. The direction of the gamma rays changes. Suggest an explanation for the change in direction. You may draw a diagram to illustrate your answer.

> Higher

....

....

....

....

(2 marks)

 5 A ray of light enters a block of glass which has a refractive index of 1.5. The ray travelling through the air hits the glass with an angle of incidence of 40°. Show that the angle of refraction is just over 25°.

> Higher

Snell's Law: $\dfrac{\sin i}{\sin r}$ = refractive index

> You need a calculator with trigonometric functions to do this calculation.

Angle of reflection ....

(2 marks)

Critical angle

1 Complete the diagram showing what happens to most of the light ray when it hits the boundary at an angle of incidence:

  a) less than the critical angle. Label this line A.

  b) more than the critical angle. Label this line B.

> Don't forget to label your lines A and B.

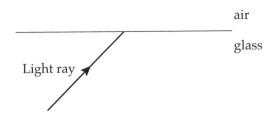

(2 marks)

2 Archaeologists find pieces of glass in a dig. To compare it with modern glass they need to measure the refractive index of the glass. A sample is cut into a rectangular shape. Suggest a method of determining the refractive index of the glass.

Apparatus needed: source of light ray and protractor ....

How it is done: ....

....

Measurements made: ....

(3 marks)

3 Snell's Law: $\dfrac{\sin i}{\sin r} = $ refractive index

> You will need a scientific calculator for these questions.

  a) The sparkle of a cut diamond is partly due to the very high refractive index of 2.4. Calculate the critical angle for diamond.

> Remember that in calculations using Snell's Law involving the critical angle, angle i is the angle the ray in the air makes with the normal and angle r is the critical angle, c. So $\sin c = \dfrac{1}{\text{refractive index}}$

Critical angle..

(2 marks)

  b) Polycarbonate is a transparent polymer used in safety glasses. The critical angle in polycarbonate is 39.12°. Calculate the refractive index of polycarbonate.

Refractive index ....

(2 marks)

Had a go ☐ Nearly there ☐ Nailed it! ☐

Using reflection and refraction

F-D

EXAM ALERT

1 The diagram shows a length of optical fibre. Complete the diagram to show how the ray of light travels along the fibre.

Light ray

Students have struggled with exam questions similar to this – **be prepared!** ResultsPlus

If you are asked to draw a ray diagram it is best to use a ruler.

(2 marks)

D-B

2 **a)** Explain why light will travel along an optical fibre without escaping from the sides.

Remember that optical fibres are strands of solid glass.

....

....

(2 marks)

b) Endoscopes are used by doctors to examine inside the digestive system of people. The endoscope has a long thin cable that can be pushed down the patient's gullet. The doctor looks through an eyepiece at one end of the cable. Explain how the doctor is able to see into the patient's stomach.

....

....

(2 marks)

D-B

Guided

3 Ultrasound is used to provide doctors with images of organs in the body. Describe how the image is obtained.

A probe sends .... into the body.

At the boundary between different tissues the waves are..

The detector records the ....

(3 marks)

B-A*

Higher

4 A new use of ultrasound in medicine is to kill cancer cells. High-intensity ultrasound waves are directed in a narrow beam at the tumours that lie inside the body. The cancer cells become hot and die.

You may need to refer to earlier pages to answer these questions.

a) Explain why the cancer cells become hot.

....

....

(2 marks)

b) State why high-intensity ultrasound is necessary.

....

(1 mark)

c) State why the ultrasound waves must be confined to a narrow beam.

....

(1 mark)

Physics extended writing 1

Modern telephone, cable television and broadband internet connections to our homes rely on optical fibres which are made of glass. Optical fibres are also used in endoscopes used by doctors to investigate inside the body and by rescuers searching under fallen buildings.

Explain how light travels through flexible optical fibres and how this is used in communications and endoscopes.

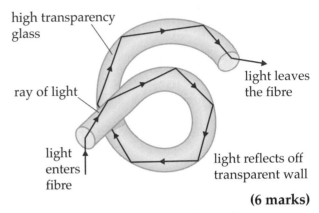

high transparency glass

ray of light

light enters fibre

light leaves the fibre

light reflects off transparent wall

(6 marks)

> You will be more successful in extended writing questions if you plan your answer before you start writing. In this question you need to think about how light is trapped inside an optical fibre.
>
> Some questions to think about are:
> - What happens when light travelling through glass reaches a boundary with the air?
> - What is meant by the critical angle and total internal reflection?
> - What happens to light when it travels down an optical fibre?
> - How are optical fibres used in communications and endoscopes?

....

....

....

....

....

....

....

....

....

....

....

....

....

....

....

X-rays

F-C **1 a)** The following sentences describe how X-rays are
generated in an X-ray machine like the one in the diagram.
The sentenes are not in order. Write down the letters at the
start of the sentences in the correct order.

A Electrons are given off from the hot cathode.

B The metal anode gives off X-rays.

C The cathode filament is heated.

D The electrons hit the metal anode.

E The electrons accelerate towards the metal anode.

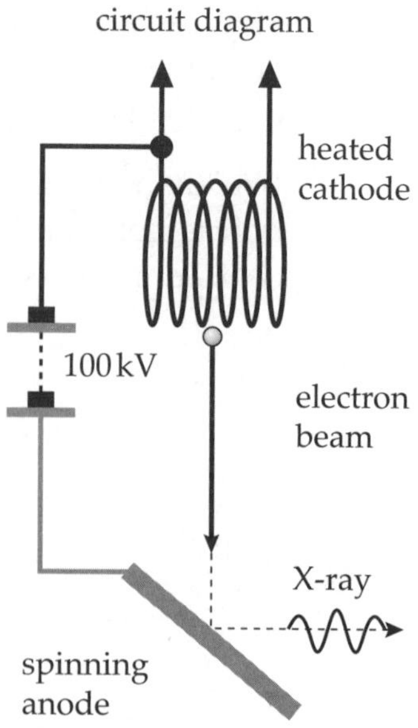

Answer ....

(3 marks)

D-A **b)** Look at the sentences above.
Write down the letter of the sentence that:

i) shows the electrons and anode have opposite charges: ....   **(1 mark)**

ii) describes thermionic emission: ....   **(1 mark)**

Guided **c)** Explain what happens when the potential difference between the anode and cathode is
increased.

With a higher potential on the anode the electrons hit the anode with ....

the X-rays produced have ....

(2 marks)

d) The X-ray tube has to be evacuated.
Describe what would happen if air
was allowed into the tube.

> Evacuated means 'to make a vacuum'. Think about the
> difference between a vacuum and a container full of air.

....

....

(2 marks)

e) Describe the similarity between the beam of electrons leaving the cathode in the X-ray
tube and a current in a metal wire.

....

....

(2 marks)

B-A* **2** There are two main controls on an X-ray machine – the cathode temperature control and the
Higher   potential difference control. Explain which control will vary the intensity of the X-ray beam.

....

....

....

(3 marks)

circuit diagram

heated
cathode

100 kV

electron
beam

X-ray

spinning
anode

X-ray calculations

You will need these formulae to answer these questions:

Current = number of particles per second × charge on each particle $I = N \times q$

Kinetic energy = electronic charge × potential difference $KE = \frac{1}{2} \times m \times v^2 = e \times V$

The charge on an electron = 1.6×10^{-19} C

> This figure is used in all of the questions below.

1 a) The cathode of an X-ray machine produces a beam of electrons at the rate of 3.125×10^{18} electrons per second. Calculate the current. State the unit.

> Make sure you understand how to use the equation $I = N \times q$

> Students have struggled with exam questions similar to this
> – **be prepared!**   ResultsPlus

$$I = N \times 1.6 \times 10^{-19} \, C =$$

Current.. unit ....

(3 marks)

b) The accelerating voltage in the X-ray machine is 80 kV. Calculate the energy of the electrons when they hit the anode. State the unit.

$$KE = 1.6 \times 10^{-19} \, C \times V =$$

Energy.. unit ....

(3 marks)

2 Electron guns as used in X-ray machines are also used in 'old style' cathode ray tube television sets and computer displays.

a) The cathode ray current of a TV set is 0.4 A. Calculate the number of electrons being emitted every second.

Number .... electrons/second

(3 marks)

b) The accelerating voltage in a TV set is 40 kV. Calculate the velocity of the electrons in the electron beam. The mass of an electron is 9.11×10^{-31} kg.

> You can check if your answer is correct – it should be about $\frac{1}{3}$ of the speed of light!

Velocity.. m/s

(3 marks)

Using X-rays

F-D 1 In X-ray photographs of the human body the bones show up white and other tissues in shades of grey. Which of the following is the best explanation for this?
Put a cross in the box (☒) next to your answer.

☐ **A** Bone is denser than other tissue and absorbs more of the X-rays.

☐ **B** Bone is denser than other tissue and absorbs less of the X-rays.

☐ **C** Bone is less dense than other tissue and absorbs more of the X-rays.

☐ **D** Bone is less dense than other tissue and absorbs less of the X-rays.

(1 mark)

D-B 2 When a dentist takes an X-ray photograph of a patient's teeth she places the X-ray source as close to the teeth as possible. Explain why this is done.

> **Guided**

When the source is close the X-rays will not.. too much

so the .... is high enough to give a sharp picture.

(2 marks)

D-B 3 Clarence Dally was an assistant to the American inventor Thomas Edison. From 1895 to 1900 he worked on developing the X-ray fluoroscope. He frequently put his hands in the machines to test them. Dally developed cancers that required his arms to be amputated and he died in 1904. Suggest reasons why we use X-rays today in CAT scans and fluoroscopes in medicine and airport security in spite of what happened to Clarence Dally.

....

....

(2 marks)

B-A* 4 **a)** Describe how radiologists would use X-rays to obtain the following data on patients.

> **Higher**

   i) A picture of a slice through the patient's brain.

   ....

   ....

   ....

(3 marks)

   ii) A video of blood flowing through the heart.

   ....

   ....

   ....

(3 marks)

   b) The intensity of X-rays 3 cm from a source is 100 W/m². State the intensity of the X-rays 6 cm from the source.

Intensity.. W/m²

(2 marks)

ECGs and pacemakers

 1 A pulse oximeter is an instrument that is clipped to a patient's finger or ear lobe. Which of the following are the types of radiation used in the pulse oximeter? Put a cross in the box (☒) next to your answer.

☐ **A**   red light and blue light   ☐ **B**   blue light and ultraviolet light

☐ **C**   red light and infrared   ☐ **D**   blue light and infrared

(1 mark)

 2 The diagram shows the ECG of a person at rest.

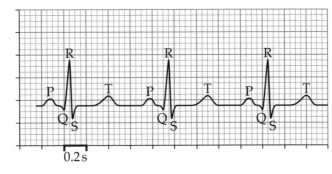

a) State how the ECG might change if the person was exercising.

....

(1 mark)

b) Pacemakers are fitted to people who have some types of heart problem. Describe what a pacemaker does.

....

....

(2 marks)

 3 A pulse oximeter emits two types of light and detects the amount of each that passes through the blood. Explain how this gives the percentage of oxygen in the blood.

The oximeter compares amounts of .... and

.... that are absorbed, because blood containing

oxygen absorbs different amounts of each type of light compared with blood

without ....   **(2 marks)**

 4 Calculate the heart rate in beats per minute from the ECG in question 2 above.

You will need the equation: frequency $= \dfrac{1}{\text{time period}}$, $f = \dfrac{1}{T}$

> Mark points on the graph to show where you are taking your measurements.

Heart rate ....

(3 marks)

Properties of radiation

F-D **1 a)** Draw a line from each type of radiation emitted from atoms to the descriptions of what the radiation is made up of.

Radiation

| alpha |
| beta |
| positron |
| gamma |
| neutron |

Description

- slow moving protons
- particles with positive charge and same mass as electron
- particles with same mass as proton but no charge
- particles made up of two protons and two neutrons
- fast-moving electrons
- electromagnetic radiation

(5 marks)

b) Different types of radiation can penetrate different distances through materials. Put alpha, beta and gamma radiation in order of penetrating power, starting with the most penetrating.

> Think about the materials needed to stop each type of radiation.

....

(1 mark)

F-C **2 a)** Atoms are made up of protons, neutrons and electrons. Explain why atoms are neutral.

....

....

(2 marks)

b) An atom of nitrogen is made up of 7 electrons, 7 protons and 7 neutrons. Explain why the mass (nucleon) number of the atom is 14 and not 21.

....

....

(2 marks)

D-C **3** $^8_3$Li is an isotope of lithium with 3 protons and 5 neutrons in its nucleus. The nucleus is unstable and gives out beta radiation when it decays.

Guided

a) Describe what happens in the nucleus to give out a beta particle.

A .... in the nucleus becomes a .... and a beta particle.

(2 marks)

b) Complete the table below for the changes that take place when the atoms emit radiation.

Element	Proton number	Nucleon number	Radiation emitted	New proton number	New nucleon number
lithium	3	8	beta		
uranium	92	238		90	
americium	95	239	gamma		

(3 marks)

Balancing nuclear equations

Higher This whole page covers Higher material.

 1 Beta decay is the emission by an atom of an electron ($\beta-$ decay) or a positron ($\beta+$ decay). Compare the processes by which electrons and positrons are emitted from atoms.

....

....

(2 marks)

 2 Identify the type of radiation emitted in the following nuclear reactions.

> The sum of the atomic numbers on the right must equal the atomic number on the left, and the same for the mass numbers.

a) $^{224}_{86}\text{Rn} \rightarrow ^{220}_{84}\text{Po} +$ ....

(1 mark)

b) $^{31}_{16}\text{S} \rightarrow ^{31}_{15}\text{P} +$ ....

(1 mark)

 3 Complete the following nuclear reaction equations by filling in the atomic and mass numbers of the products of the decay.

Guided

> The guided example shows the calculations that give the answer. These workings are not required in your answers.

a) $^{131}_{53}\text{I} \rightarrow ^{131}_{54}\text{Xe} + ^{0}_{-1}\text{e}$

(2 marks)

b) $^{239}_{94}\text{Pu} \rightarrow \quad \text{U} + ^{4}_{2}\text{He}$

(2 marks)

c) $^{13}_{4}\text{Be} \rightarrow \quad \text{Be} + ^{1}_{0}\text{n}$

(2 marks)

B-A* **4** $^{99}_{43}\text{Tc}$ is an isotope of technetium used for gamma ray imaging of tumours in organs of the body. Tc-99 is obtained by the decay of $^{99}_{42}\text{Mo}$.

a) Write an equation for the nuclear reaction.

....

(1 mark)

b) Identify the type of radiation emitted by Mo-99 when it forms Tc-99.

....

(1 mark)

The nuclear stability curve

Higher This whole page covers Higher material.

C-B 1   The graph shows the isotopes of elements.

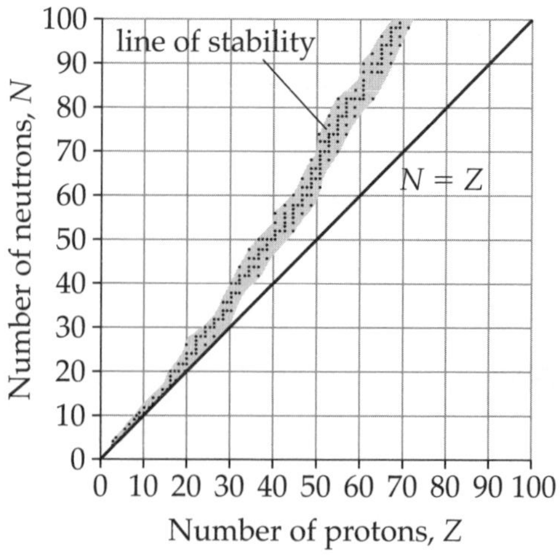

a) Which of the following statements is correct?
Put a cross (☒) in the box next to your answer.
The graph shows that:

  ☐ **A**   All stable isotopes have equal numbers of neutrons and protons.

  ☐ **B**   As the number of protons of an element increases the number of neutrons in its stable isotope becomes less than the number of protons.

  ☐ **C**   Elements with a low mass have more unstable isotopes.

  ☐ **D**   Stable isotopes of elements with a small number of protons have the same or nearly the same number of neutrons as protons.

(1 mark)

b) Use the graph to determine the number of neutrons in the stable isotope of zirconium ($Z = 40$).

....

(1 mark)

C-B 2   a) Explain which type of radioactive decay the isotope radium-224 ($Z = 88$) is most likely to undergo.

> **Guided**

    The number of protons in radium-224 is greater than ....

    so it undergoes ....

(2 marks)

b) The stable isotope of oxygen ($Z = 8$) is oxygen-16. Explain which type of radioactive decay the isotope oxygen-13 undergoes.

> Work out which side of the stable isotope line oxygen-13 is on.

....

....

(2 marks)

B-A* 3   The most stable isotope of bismuth ($Z = 83$) is bismuth-209. Bismuth-212 can decay either by emitting an electron or an alpha particle. Explain how each of these types of decay produces a more stable nucleus.

....

....

....

(3 marks)

Quarks

Higher This whole page covers Higher material.

There are two pages on Quarks in the Revision Guide to help you understand this topic, but there is only one page of questions here. On page 86 there is a page of summary questions covering all the ideas in this topic.

EXAM ALERT

Make sure you know what happens in each type of decay.

Students have struggled with exam questions similar to this – **be prepared!** ResultsPlus

C-B 1 a) Which of the following represents the type of quarks that make up a proton? Put a cross (☒) in the box next to your answer.

☐ **A** up, up, up   ☐ **B** up, up, down

☐ **C** up, down, down   ☐ **D** down, down, down

(1 mark)

Guided b) Explain why a neutron has no charge.

A neutron is made up of two .... quarks, each with a charge of .... and

one .... quark with a charge of .... making a total of zero.

(4 marks)

c) Explain why protons and neutrons have a mass number of 1.

....

....

(2 marks)

C-A 2 In a nucleus undergoing radioactive decay, particle A changes to particle B, giving off a particle X.

particle A   →   particle B   +   X
(quarks: up down down)   (quarks: up up down)

a) State the change that has occurred in the quarks.

....

(1 mark)

b) Name the type of decay this equation represents

....

(1 mark)

B-A* 3 Explain in terms of changes to the quarks how a nucleus decays by releasing a positron.

Remember that quarks do not exist on their own, so you must say in which particle the change occurs.

....

....

....

(3 marks)

Dangers of ionising radiation

F-C

Guided

1 Radioactive material producing ionising radiation is used in many branches of medicine.

a) List three precautions that are taken to reduce the amount of radiation that medical workers receive.

1. The dose of ionising radiation given to the patient is kept ....

2. ....

3. ....

(3 marks)

b) Explain why medical personnel working with radioactive sources carry instruments that measure how much ionising radiation they receive.

....

....

(2 marks)

C-A*

Higher

2 The explosion at Chernobyl in 1986 was the worst accident ever to take place in a nuclear power station. The fire at the reactor spread radioactive material across the Earth. 237 of the emergency workers suffered from radiation sickness, and 31 of them died. Estimates of the number of people around the world who will die of cancers caused by the radiation vary from 4000 to 1 000 000.

> The radiation dosage close to the reactor was many times higher than that produced by the material that fell on other countries.

a) Explain how ionising radiation affected the emergency workers at Chernobyl.

....

....

(2 marks)

b) Explain how ionising radiation causes cancer.

....

....

(2 marks)

c) In the two years after the Chernobyl accident there was a doubling of the number of birth defects in babies born in countries close to Chernobyl. Describe how ionising radiation causes birth defects.

....

....

(2 marks)

d) Suggest reasons why there is a definite number of deaths from radiation sickness but the number of deaths from cancer is only an estimate.

> Remember that the estimate includes people who have not yet died.

....

....

(2 marks)

Radiation in hospitals

 1 The jumbled sentences below describe how a PET scan is carried out. Write down the letters at the start of each sentence in the correct order.

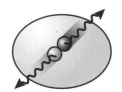

     A The gamma rays are detected on opposite sides of the body and image is built up.

     B The tracer collects in the tissues being examined and emits positrons.

     C The patient is injected with tracer molecules containing atoms of a radioactive isotope.

     D Positrons annihilate electrons and give out two gamma rays.

     E The gamma rays travel in opposite directions.

A positron annihilates an electron and gives out two gamma rays.

Answer ....

(3 marks)

D-B **2** The following are radioactive isotopes used in medicine.

> **Guided**

     A cobalt-60, a gamma emitter   **B** iodine-131, a beta emitter

     C uranium-238, an alpha emitter   **D** carbon-11, a positron emitter

> You need to recall the properties of the different types of radiation covered earlier in this topic.

Explain which isotope you would use for the following purposes.

     a) Concentrates in particular organs and kills the surrounding cancer cells.

        Isotope: .... because the .... can travel a few centimetres ....

(2 marks)

     b) Held in a sealed container outside the body to direct ionising radiation to the affected part.

        Isotope: .... because the .... can travel into the body some distance ....

(2 marks)

C-A* **3** The use of radioactive isotopes in medicine raises ethical questions because ionising radiation can kill healthy cells as well as cancerous cells. Suggest reasons for the following.

> **Higher**

> In questions of ethics you must decide whether the benefits outweigh the costs of an action.

     a) The level of ionising radiation used in treating cancers is higher than in diagnosis.

        ....

        ....

(2 marks)

     b) Strontium-90 is used in palliative care for patients with bone cancer although it doesn't cure the cancer and may destroy healthy cells.

        ....

        ....

(2 marks)

Radioactivity summary

F-C **1** Iodine-131 is a beta radiation emitter used to treat thyroid cancer.

a) Beta rays are electrons. State the mass and the charge of an electron compared with a proton.

> Take the mass of a proton to be 1.

....

(2 marks)

b) Describe the change that takes place in the nucleus of iodine-131 to emit a beta particle.

....

....

(2 marks)

Guided **c)** Explain why medical workers make sure iodine-131 is kept in a metal box until used.

The beta particles are .... by the metal box so the medical

workers .... from the harm that the beta particles can cause.

(2 marks)

C-A*

Higher

2 To carry out a PET scan, a patient is injected with a tracer such as a glucose molecule in which a fluorine-18 atom replaces an oxygen and a hydrogen atom. The glucose substitute is taken up by cells like normal glucose. Fluorine-18 is a positron emitter with a half-life of 110 minutes.

a) Where in the graph of neutron number against proton number will the fluorine-18 isotope be found?
Put a cross in the box (☒) next to your answer.

> A copy of the graph is on page 82.

☐ **A** On the curve for stable isotopes.

☐ **B** Above the curve for stable isotopes.

☐ **C** Below the curve for stable isotopes.

☐ **D** It does not appear because it is not stable.

(1 mark)

b) Complete the balanced nuclear equation below for the positron emission of fluorine-18.

$$^{18}_{9}F \rightarrow \boxed{}\!\!\boxed{} O + ^{0}_{1}\beta^{+}$$

(2 marks)

c) Describe the change that takes place to a quark in the fluorine-18 nucleus when it gives out a positron.

....

(2 marks)

d) Explain why fluorine-18 has to be made close to where the scan takes place.

....

....

(2 marks)

Physics extended writing 2

CAT scans are frequently used in hospitals to provide an image through a patient's body. X-rays are harmful. Explain how the image in a CAT scan is obtained and how the risk of harm to the patient and medical staff is kept as low as possible.

(6 marks)

You will be more successful in extended writing questions if you plan your answer before you start writing. In this question you need to think about the properties of X-rays.

Some questions to think about are:

- What happens to an X-ray beam when it is emitted from the tube?
- What happens to X-rays when they pass through the body?
- What precautions ensure the lowest possible dose of X-rays to the patient?

Remember that in an 'explain' question you are expected to say what happens *and why*. Try to use scientific terms in your answer.

....
....
....
....
....
....
....
....
....
....
....
....
....
....
....
....
....
....
....

Physics extended writing 3

Higher Various types of ionising radiation are used in medical diagnosis and treatment. The radiation is produced by the decay of many different radioactive isotopes.

Explain the processes that form the different types of ionising radiation.

(6 marks)

You will be more successful in extended writing questions if you plan your answer before you start writing. In this question you need to think about how different radioactive isotopes decay.

Some questions to think about are:

- What types of radiation are produced by radioactive decay?
- How are these types of radiation formed?
- Why do different radioisotopes produce different types of radiation?

Try to use scientific terms in your answer.

Remember that in an 'explain' question you are expected to say *what* happens and *why*.

....
....
....
....
....
....
....
....
....
....
....
....
....
....
....
....
....
....
....

Particle accelerators

F-C   **1**   A centripetal force makes a particle in a particle accelerator move in a circle around a point. What happens when the centripetal force is removed? Put a cross in the box (☒) next to your answer.

> Remember that a centripetal force acts towards the centre of a circle.

     ☐ **A**   The particle moves towards the central point.

     ☐ **B**   The particle moves away from the central point.

     ☐ **C**   The particle moves in a straight line at a tangent to the circle.

     ☐ **D**   The particle stops moving.

                                                     (1 mark)

E-C   **2**   Name the part of the cyclotron that

     a)   increases the speed of the charged particles

        ....

        ....

                                (1 mark)

     b)   makes the charged particles move in a circle

        ....

        ....

                                (1 mark)

high-frequency alternating voltage

electromagnet

N-pole

radioactive source

accelerated particles

S-pole

D-B   **3**   Cyclotrons can produce radioactive isotopes for medical use. Explain how this is done.

   Guided   .... produced by the cyclotron collide with a target of

   ....

                                                                     (2 marks)

C-A*   **4**   The Large Hadron Collider cost over £6 billion to build. Suggest reasons that justify this expense.

   Higher

> Thousands of scientists work at the LHC.

   ....

   ....

                                                                     (2 marks)

Collisions

 1 A proton smashes into the nucleus of a strontium atom. The collision is inelastic. What does this tell you about whether the kinetic energy and momentum are conserved in the reaction? Put a cross in the box (☒) next to the answer you choose.

> The most important word in the question is 'inelastic'.

Answer box		Kinetic energy conserved	Momentum conserved
	A	Yes	Yes
	B	No	No
	C	Yes	No
	D	No	Yes

(1 mark)

 2 A runaway railway truck collides with another truck with the same mass on a level track. It is an elastic collision. The first truck stops. Explain what happens to the second truck.

The second truck moves with the same kinetic energy and velocity as ....

because energy and momentum are ....

(2 marks)

3 Look at the diagram. Both trolleys have the same mass and speed.

Before collision

 velocity v 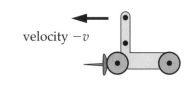 velocity $-v$

> You should remember which quantities are vectors and which are not.

After collision

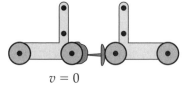

$v = 0$

a) Explain why the velocity of the right-hand trolley has a minus sign in front of the letter v.

....

....

(2 marks)

b) Describe what happens to the kinetic energy of the trolleys in the collision.

> Compare the total kinetic energy of the two trolleys before and after the collision.

....

....

(2 marks)

c) Describe what happens to the momentum of the trolleys in the collision.

....

....

(2 marks)

d) Decide whether the collision is elastic or inelastic and explain your decision.

....

....

(2 marks)

Calculations in collisions

Higher This whole page covers Higher material.

You will need to use these formulae:

Kinetic energy $= \frac{1}{2} \times$ mass $\times$ velocity^2 $KE = \frac{1}{2}mv^2$

Momentum = mass $\times$ velocity $p = m \times v$

1 A ball with a mass of 0.25 kg travelling at 4 m/s collides with a ball of mass of 0.5 kg that is not moving. The 0.25 kg ball stops and the heavier ball starts to move.

a) Calculate the velocity of the heavier ball after the collision.

Total momentum before collision = momentum of small ball = m X v = ...

Total momentum after collision = ...

Velocity of larger ball = momentum/mass = ...

Velocity .... m/s

(3 marks)

b) Show that the collision is inelastic.

> Calculate the kinetic energy of the balls before and after the collision.

....

(3 marks)

2 Protons and neutrons each have a mass of one unit. A proton travelling at 1×10^8 m/s collides with a stationary nucleus of the isotope lithium-7. The proton joins with the lithium nucleus to form a nucleus of isotope beryllium-8. Calculate the velocity of the beryllium nucleus.

> You normally need a mass in kilograms to use the equations for momentum and kinetic energy. However, you can use relative masses as all the masses involved are given as relative masses. When you use the relative masses of the proton and neutron in the calculation of the momentum you can omit units.

Velocity .... m/s

(3 marks)

3 A ball of mass 1 kg and velocity 4 m/s collides with a ball of mass 2 kg and velocity -2 m/s in an elastic collision. Calculate the velocity of both balls after the collision.

Answer ....

(5 marks)

Had a go ☐ Nearly there ☐ Nailed it! ☐

Electron–positron annihilation

E-C 1 The diagram shows what happens when a positron meets an electron.

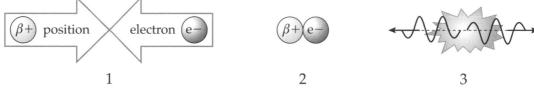

$$1 \qquad\qquad 2 \qquad\qquad 3$$

a) Which of the following statements correctly explains the conservation of charge when a positron annihilates an electron? Put a cross in the box (☒) next to your answer.

☐ **A** The sum of the charges before and after the collision is zero.

☐ **B** The positron has the same charge as the electron.

☐ **C** The gamma rays formed by the collision have the same charges as the positron and electron.

☐ **D** Both a position and an electron have a negative charge. **(1 mark)**

b) In a PET scan a tracer containing a radioisotope is injected into the patient. Which of the following happens next? Put a cross in the box (☒) next to your answer.

> You may have to look back to earlier work for a reminder about radioisotopes and tracers.

☐ **A** The tracer emits an electron.

☐ **B** The tracer emits a positron.

☐ **C** The tracer emits a gamma ray.

☐ **D** The tracer emits a positron and an electron. **(1 mark)**

C-A 2 **a)** When a positron and electron collide, gamma rays are formed that have more energy than the positron and electron. Explain how this obeys the law of conservation of energy.

Higher

Guided The mass of the positron and electron is converted into the energy ....

.... so mass-energy is ....

(2 marks)

b) Explain how momentum is conserved in positron–electron annihilation.

....

....

....

(3 marks)

B-A* 3 A positron annihilates an electron.
Calculate the energy released in this process.

Higher

You will need the formula $E = mc^2$

> Be careful doing calculations with figures in standard form. To square a number, add the powers of 10 together.

Mass of an electron = mass of a positron = 9×10^{-31} kg; the speed of light = 3×10^8 m/s

Energy released ....J

(3 marks)

Kinetic theory

F-D

Guided

1 a) Match each of the states of matter to the statements about the kinetic theory. There may be two statements linked to a single state of matter.

States of matter   Statements of the kinetic theory

Solids		Particles are far apart and move quickly.
Liquids		Particles are held closely by strong forces.
Gases		Particles are close together but can move past each other.
		Particles can vibrate but cannot move freely.

(3 marks)

b) If you squeeze an inflated balloon you can feel the pressure of the gas pushing back against your hands. Describe how the particles of the gas exert pressure on the surface of the balloon.

The particles in the gas ....

with the surface of ....

(2 marks)

E-C

2 A researcher measured the volume of a gas at different temperatures and drew the graph shown.

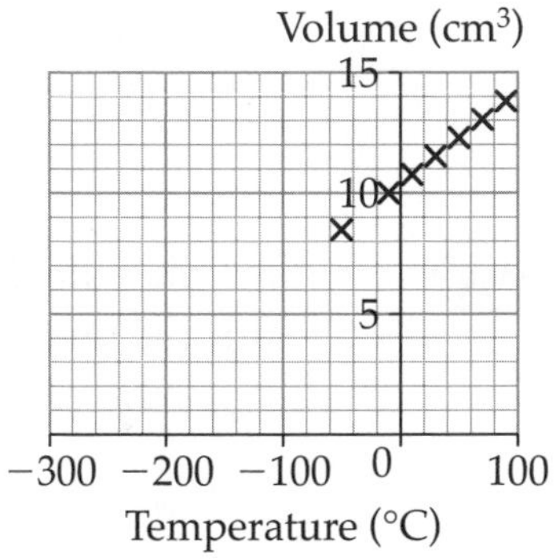

Volume (cm³)

a) State the name given to the temperature at which a line through the points meets the temperature axis.

....

....

(1 mark)

−300 −200 −100   0   100
Temperature (°C)

b) Describe what happens to the particles of a gas as the temperature is decreased to the temperature mentioned in **a)**.

> Just focus on individual particles and ignore changes of state.

....

....

(2 marks)

D-B

3 a) A gas turns to a liquid at a temperature of −78°C. Calculate what this temperature is on the Kelvin scale. (0 K is −273°C)

Temperature ....

(2 marks)

b) Describe the shape you would expect for a graph of temperature (in Kelvin) against the average kinetic energy of the particles in a gas.

....

....

(2 marks)

Ideal gas equations 1

You will find these formulae useful in the following questions:

The relationship between temperature and volume of a gas: $V_1 = \dfrac{V_2 T_1}{T_2}$

The relationship between pressure and volume of a gas: $P_1 V_1 = P_2 V_2$

> Pressures are usually in pascals and volumes in m^3, but in these equations you can use other units (such as atmospheres or N/cm^2 for pressure, and cm^3 or litres for volume) as long as both pressures are in the same units, and both volumes are in the same units. Don't forget to use K for temperature.

1 A student set up the apparatus shown and wrote down the reading on the ruler when the water was at different temperatures.

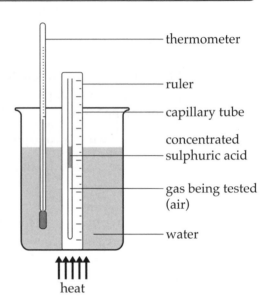

thermometer

ruler

capillary tube

concentrated sulphuric acid

gas being tested (air)

water

heat

a) What quantities were being kept constant during this investigation?
Put a cross in the box (☒) next to your answer.

☐ **A** volume and temperature

☐ **B** temperature and mass

☐ **C** mass and pressure

☐ **D** pressure and volume

(1 mark)

> **Guided**

b) Calculate the volume of gas in the tube at 77 °C if the bubble of gas in the tube had a volume of 0.6 cm³ at 27 °C.

> Remember that 0 °C is 273 K

$T_1 = 77\,°C = $ .... K, $T_2 = 27\,°C = $ .... K, $V_2 = 0.6\ cm^3$

$V_1 = V_2 T_1 / T_2 = $

Volume .... cm³

(3 marks)

2 a) A hospital patient is receiving oxygen at normal atmospheric pressure at a rate of 2 litres/minute from a cylinder. Explain why the cylinder has to be able to withstand very high pressures.

....

....

(2 marks)

b) The patient is supplied with a cylinder with a volume of 10 litres, which contains oxygen at a pressure of 200 atm. Show that the cylinder should last up to 16 hours.

> Make sure you put the figures in the correct place in the formula.

....

(3 marks)

Ideal gas equations 2

Higher This whole page covers Higher material.

1 A mixture of nitrous oxide and oxygen gases is used as an anaesthetic by dentists. 640 litres of the gas at normal atmospheric pressure (100 000 Pa) can be stored in a cylinder with an internal volume of 8 litres.

a) Calculate the pressure needed to compress the gas into the cylinder.

> You need to rearrange the formula given at the top of the previous page.

Pressure .... Pa

(3 marks)

b) The gas is released from the cylinder at a rate of 2 litres per minute. Calculate the time the gas supply will last.

Time .... minutes

(2 marks)

> For the following questions you will need the formula: $\dfrac{P_1 V_1}{T_1} = \dfrac{P_2 V_2}{T_2}$

2 Climbers usually need an oxygen supply to reach the top of Mount Everest, but some have tried without. An adult male exercising can take in $0.0005\,\text{m}^3$ of air in each breath.

a) Calculate the volume of air at the top of Everest that would contain the same mass of air as the climber can inhale in one breath at sea level.

At the top of Everest: air pressure = 34 000 Pa, temperature = −26 °C

Sea level: air pressure = 100 000 Pa, temperature = 17°C

$P_1 = 100\,000\,\text{Pa}, \; V_1 = 0.0005\,\text{m}^3, \; T_1 = 17\,°\text{C} = \text{..........} \; \text{K}$

$P_2 = 34\,000\,\text{Pa}, \; T_2 = -26\,°\text{C} = \text{..........} \; \text{K}$

$V_2 =$

(3 marks)

b) Explain why climbers usually need to carry bottled oxygen for the climb.

> Remember that lung capacity is fixed.

....

....

(2 marks)

3 A hot air balloon has a fully inflated volume of $2800\,\text{m}^3$. $2400\,\text{m}^3$ of air at normal air pressure (101 000 Pa) and at 12 °C is heated and blown into the deflated balloon. When the balloon is fully inflated, the pressure reaches 104 000 Pa. Calculate the temperature of the air in the fully inflated balloon.

Temperature .... K

(3 marks)

Physics extended writing 4

Cyclotrons were first used by physicists exploring the particles that make up matter. Increasingly they are used in hospitals as a way of making radioisotopes.

Explain how a cyclotron produces high-velocity particles that can be used to make radioisotopes and why it is useful to have one in a hospital.

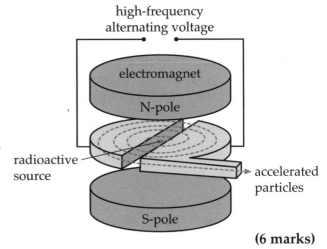

high-frequency alternating voltage

electromagnet

N-pole

radioactive source

accelerated particles

S-pole

(6 marks)

You will be more successful in extended writing questions if you plan your answer before you start writing. In this question you need to think about how cyclotrons work and their uses.

Some questions to think about are:

- What are the parts of a cyclotron and how do they affect the particles?
- What happens when the particles from the cyclotron hit a target?
- Why it is useful for hospitals to have the means of manufacturing radioisotopes?
- How does the momentum and energy of a particle change in the cyclotron and in collisions?

Remember that in an 'explain' question you are expected to say *what* happens and *why*. Try to use scientific terms in your answer.

....

....

....

....

....

....

....

....

....

....

....

....

....

....

....

Physics extended writing 5

The kinetic theory predicts that there is an absolute zero of temperature at which gases exert no pressure. Absolute zero cannot be measured directly but instead is found by extending a graph of pressure against temperature of a gas measured at higher temperatures.

Describe how the kinetic theory explains the properties of gases and the absolute zero of temperature.

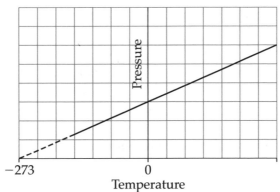

Change in pressure of a fixed volume of gas with temperature

(6 marks)

You will be more successful in extended writing questions if you plan your answer before you start writing. In this question you need to think about how the kinetic theory explains the properties of gases

Some questions to think about are:

- What is the kinetic theory?
- In what ways are gases different to solids and liquids?
- How does the kinetic theory explain the pressure a gas exerts at different temperatures?
- Why is it impossible to measure absolute zero directly?

....

....

....

....

....

....

....

....

....

....

....

....

....

....

Biology practice exam paper (allow one hour)

Edexcel publishes official Sample Assessment Material on its website. This practice exam paper has been written to help you practise what you have learned and may not be representative of a real exam paper.

This paper is a Higher tier paper. Sample Foundation tier papers can be found on the Edexcel website.

Functions of the kidney

1   During metabolism, the human body forms waste products that build up in the blood. The body is able to remove these waste products.

a)   Urea is one waste product of metabolism.

  i)   Name one other waste product of metabolism.

  ....

  (1 mark)

  ii)   State where and how urea is made.

  ....

  ....

  (2 marks)

b)   Many nephrons are found in each kidney.

  i)   The diagram shows a nephron from a healthy person. Place an X on the diagram at the site where the glomerulus would be found and a Y at the loop of Henle.

  (2 marks)

  ii)   Suggest how the blood and urine leaving this nephron may change if selective reabsorption of glucose did not occur.

  ....

  (1 mark)

c)   Ethanol is believed to suppress the release of ADH from the pituitary gland. Explain how ethanol intake would affect the water content of the blood.

  ....

  ....

  ....

  (3 marks)

d)   Describe the route urine follows to leave the body after it has left the nephron.

  ....

  ....

  (2 marks)

  (Total for Question 1 = 11 marks)

Immune system

2   Humans can be harmed by certain microorganisms. Humans have mechanisms to defend
themselves.

Person P catches a bacterial infection and becomes unwell. One week later he has recovered.
Person Q was immunised against this bacterial infection.

a)   Complete the table below by writing in the name of the appropriate cell.

Description	Appropriate cell
Cell producing antibody in person P	
Cell that produces antigen	
Cell that keeps person Q immune	

(3 marks)

b)   Suggest **one** risk to person Q when they are immunised against a bacterial infection.

....

(1 mark)

c)   Antibodies are part of our immune system. They are very specific for their antigens.
Scientists have been able to produce special antibodies from hybridoma cells to target
cancer cells. Name the protein.

....

(1 mark)

d)   Explain the advantages of using these antibodies to treat cancer cells compared with
using radiotherapy.

....

....

....

(3 marks)

(Total for Question 2 = 8 marks)

Growth of microorganisms

3   In the right conditions microorganisms grow rapidly.

a)   A student tested the effect of the type of milk on microorganism growth. A blue dye called resazurin can be used to show the growth of a population of microorganisms. Resazurin is a blue dye but it becomes clear if there are lots of bacteria present.

The student placed some pasteurised milk in a test tube. She placed sterilised milk in another tube. She put the tubes in a water bath set at 36 °C and incubated them for one day. Then she added blue resazurin dye to each tube. One hour later she recorded the colour. Her results are shown in the table below.

Type of milk	Resazurin colour after one hour
Pasteurised	Clear
Sterilised	Blue

i)   Suggest why a waterbath is used.

....

(1 mark)

ii)   Suggest why the water bath was set at 36 °C for this study.

....

(1 mark)

iii)   Explain why there was a colour change for pasteurised milk but not the sterilised milk.

....

....

....

(3 marks)

iv)   The tube with sterilised milk in acted as a control. State the function of this control.

....

(1 mark)

b)   i)   The graph shows what happens to the number of bacteria in a sample of milk over time. Describe what the graph shows.

....

....

(2 marks)

ii)   Suggest what might happen to the growth curve eventually.

....

....

(2 marks)

(Total for Question 3 = 10 marks)

Plant defences

4   It is believed that a type of tree, called the Acacia, co-evolved with ants. If an elephant tries to eat the leaves, it shakes the tree and the ants swarm out of hollows they have dug in the thorns of the tree. This deters the elephants. The tree supplies food for the ants from small swellings near its leaves.

a)   i)   Suggest how the tree has changed to enable the ants to successfully live on it.

....

....

(2 marks)

ii)   Suggest the benefit for the Acacia of having ants living on it.

....

....

(2 marks)

b)   While the tree supplies food to the ants, it inhibits them from feeding on newly opened flowers by releasing a chemical. This chemical does not affect other insects. Suggest why the Acacia tree inhibits the ants from feeding on newly opened flowers.

....

....

....

(2 marks)

c)   Suggest why the Acacia tree does not inhibit the ants from feeding on old flowers.

....

....

(2 marks)

(Total for question 4 = 8 marks)

Inheritance

5   Sperm and egg are both gametes.

a)   Describe the function of a sperm.

....

....

....

(3 marks)

b)   Describe what happens to an egg to stop more than one sperm completing its function.

....

(1 mark)

c) Female sex cells always contribute an X chromosome at fertilisation and male sex cells contribute a Y chromosome. The genes carried on the X and Y chromosomes are different and can determine whether someone is colour blind or not. Explain why there are more colour blind males than females. Use a Punnett square as part of your answer.

....

....

....

....

....

....

(6 marks)

(Total for Question 5 = 10 marks)

Enzymes

6 **a)** Food can cause stains if it gets onto clothing. Biological washing powders are usually better than non-biological powders at removing food stains. Explain one way that biological washing powders work that is different to non-biological washing powders. Give examples to illustrate your answer.

....

....

....

(3 marks)

b) Milk can be treated by being passed through a tube containing immobilised lactase in beads as shown in the diagram.

Explain the functions of the beads containing immobilised lactase.

....

....

....

....

....

....

(4 marks)

Direction of milk flow

Beads containing immobilised lactase

Converted milk being collected

c) Using microorganisms for food production has many advantages over producing food from animals or plants. Describe some of these advantages.

....

....

....

....

....

....

....

....

....

....

....

....

....

....

(6 marks)

(Total for Question 6 = 13 marks)

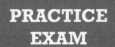
Chemistry practice exam paper (allow one hour)

Edexcel publishes official Sample Assessment Material on its website. This practice exam paper has been written to help you practise what you have learned and may not be representative of a real exam paper.

This paper is a Higher tier paper. Sample Foundation tier papers can be found on the Edexcel website.

Homologous series

1   Here are the formulae of some members of the same homologous series of compounds.

$$C_2H_4 \qquad C_4H_8 \qquad C_7H_{14} \qquad C_{10}H_{20}$$

a)   What is the general formula for molecules in this homologous series? Put a cross in the box (☒) next to your answer.

☐ **A** C_nH_n

☐ **B** C_nH_{n+2}

☐ **C** C_nH_{2n}

☐ **D** C_nH_{2n+2}

(1 mark)

b)   The second member of this homologous series has the formula C_3H_6.

i)   Give the name of this molecule.

....

(1 mark)

ii)   Draw the structure of this molecule, showing all the bonds clearly.

(2 marks)

c)   State **two** features of members of the same homologous series.

....

....

(2 marks)

d)   Vegetable oils contain molecules which are similar to molecules found in this homologous series. Explain how this vegetable oil can be made into margarine.

....

....

(2 marks)

(Total for Question 1 = 8 marks)

Measuring carbon dioxide

2 A student is reacting calcium carbonate with hydrochloric acid. He uses different masses of calcium carbonate and measures the volume of carbon dioxide that is given off. In each reaction, all the calcium carbonate reacts with the hydrochloric acid.

His results are shown in the table.

Mass of calcium carbonate in g	0.10	0.18	0.25	0.36	0.42
Volume of carbon dioxide in cm³	24	44	60	86	100

a) Which equation correctly shows the reaction between calcium carbonate and hydrochloric acid? Put a cross in the box ($\boxtimes$) next to your answer.

☐ **A** $CaCO_3 + HCl \rightarrow CaCl_2 + H_2O + CO_2$

☐ **B** $CaCO_3 + 2HCl \rightarrow CaCl_2 + H_2O + CO_2$

☐ **C** $CaCO_3 + 2HCl \rightarrow CaCl_2 + 2H_2O + CO_2$

☐ **D** $CaCO_3 + 4HCl \rightarrow CaCl_2 + 2H_2O + CO_2$

(1 mark)

b) Plot the results of his experiment on the grid below and draw a line of best fit through the points.

(3 marks)

c) i) Use your graph to predict the volume of carbon dioxide that would be made if he started with 0.30 g of calcium carbonate.

....

....

....

(1 mark)

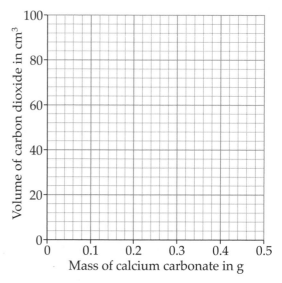

ii) Use your reading from i) to calculate the volume of carbon dioxide that would be made if he used 100 g of calcium carbonate.

....

....

(2 marks)

iii) State the significance of the number that you have calculated in ii).
(Relative formula mass of calcium carbonate = 100)

....

(1 mark)

(Total for Question 2 = 8 marks)

Testing table salt

3 Packets of table salt, sodium chloride, often have small amounts of other substances added.
These substances help keep us healthy and also stop the salt crystals from sticking together
in lumps. The diagram shows an ingredients list from a packet of table salt.

> **INGREDIENTS:**
> sodium chloride, potassium
> iodide, potassium ferrocyanide.

a) Potassium ferrocyanide contains iron (II) ions.

  i) What is the formula of an iron (II) ion? Put a cross in the box (☒) next to your
      answer.

   ☐ **A** Fe^{2+}

   ☐ **B** Fe_2^+

   ☐ **C** $2Fe^+$

   ☐ **D** $Fe\,(II)^+$

                                                                                            (1 mark)

  ii) Give the test, with its result, for the iron (II) ion.

   Test ....

   Result ....

                                                                                            (2 marks)

b) A student reads this test for the iodide ion in a text book:

   'Add $2\,cm^3$ of nitric acid to the solution being tested. Then add a few
   drops of a solution of silver nitrate ($AgNO_3$). If the test is positive, a
   yellow precipitate forms.'

  i) Name the yellow precipitate formed in a positive test for the iodide ion.

   ....

                                                                                            (1 mark)

  ii) Write the balanced equation for the reaction between silver nitrate and
      potassium iodide.

   ....

                                                                                            (2 marks)

  iii) Suggest why nitric acid is added to the solution being tested.

   ....

   ....

                                                                                            (2 marks)

c) She correctly carries out an iodide ion test on the salt. However, she does not see a
   yellow precipitate. Suggest an explanation for this.

   ....

   ....

                                                                                            (2 marks)

                                                                          (Total for Question 3 = 10 marks)

Making salts

4   A student is making the soluble salt zinc nitrate. He warms some acid in a beaker. He adds a small amount of solid zinc oxide and stirs until the solid dissolves. He continues to add zinc oxide until no more dissolves and some solid is left at the bottom of the beaker.

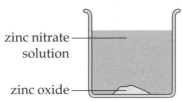

zinc nitrate solution

zinc oxide

a) i)   Name the acid that he needs to use in his experiment.

....
(1 mark)

ii) Explain why some zinc oxide is eventually left at the bottom of the beaker.

....

....
(2 marks)

iii) Name the method he should use to separate the zinc nitrate solution from the mixture in the beaker. Put a cross in the box (☒) next to your answer.

☐ **A** distillation   ☐ **B** evaporation

☐ **C** filtration   ☐ **D** fractional distillation   **(1 mark)**

b) He changes his method to make sodium sulfate. He uses sodium hydroxide solution and reacts it with sulfuric acid using the following apparatus.

i)   Explain how he would know that this reaction was complete.

....

....

....

....

....

....
(2 marks)

sulphuric acid

sodium hydroxide solution

ii) Balance the chemical equation for the reaction.

.... H_2SO_4 + ... $NaOH$ → .... Na_2SO_4 + ... H_2O

(1 mark)

iii) He uses 25.0 cm³ of sodium hydroxide solution, with a concentration of 0.50 mol dm⁻³. The sulfuric acid has a concentration of 0.20 mol dm⁻³. Calculate the volume of acid he will need to neutralise the sodium hydroxide.

....

....

....
(3 marks)

(Total for Question 4 = 10 marks)

Using electricity

5   A student is investigating electrolysis using different compounds.

a) She starts with solid lead bromide. She sets up this circuit.

The lamp does not light at first, but does after the lead bromide has been heated for a few minutes. She observes the products made at the electrodes.

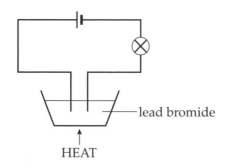

HEAT

i)   Explain why the lamp only lights after a few minutes.

....

....

(2 marks)

ii) Which row of the table shows the observations she should make at the electrodes? Put a cross in the box (☒) next to your answer.

		Negative electrode	Positive electrode
☐	A	silver liquid	red–brown fumes
☐	B	bubbles of gas	red–brown fumes
☐	C	red–brown fumes	silver liquid
☐	D	bubbles of gas	bubbles of gas

(1 mark)

iii) A reduction reaction makes the product at the negative electrode. Explain why this is a reduction reaction by using a chemical equation for the reaction occurring.

....

....

....

(3 marks)

b) She sets up a different experiment. She uses a solution of silver nitrate, $AgNO_3$, as the electrolyte.

Explain the observations that she would make in this experiment. You should give balanced equations to illustrate your answer.

pure silver — metal fork
— silver nitrate solution

....

....

....

....

....

....

....

....

....

....

....

....

....

....

(6 marks)

(Total for Question 5 = 12 marks)

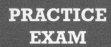

Ethanol

6   Ethanol, C_2H_5OH, is a member of the family of alcohols. It has a number of different uses.

   a) Ethanol is found in wine. If a bottle of wine is left open, oxygen from the air reacts with the ethanol.

      i) Complete the sentence. Put a cross (☒) in the box next to your answer.
In this reaction, the ethanol is

        ☐ **A** fermented

        ☐ **B** neutralised

        ☐ **C** oxidised

        ☐ **D** reduced

(1 mark)

      ii) Name the product formed in the reaction.

        ....

(1 mark)

   b) Ethanol reacts with ethanoic acid, CH_3COOH, to produce an ester.

      i) Give the name and formula of the ester formed.

        ....

        ....

(2 marks)

      ii) Describe why esters are useful.

        ....

        ....

(2 marks)

   c) Ethanol can be manufactured by two different processes. Explain why both methods are still used to make ethanol. Your answer should include details of how ethanol is made in each process.

....

....

....

....

....

....

....

....

....

....

....

(6 marks)

(Total for Question 6 = 12 marks)

Physics practice exam paper (allow one hour)

Edexcel publishes official Sample Assessment Material on its website. This practice exam paper has been written to help you practise what you have learned and may not be representative of a real exam paper.

This paper is a Higher tier paper. Sample Foundation tier papers can be found on the Edexcel website.

Particle motion

1   Clackers were once popular toys. By moving the ring up and down the pair of identical hard balls swing out equally on either side and then collide with a loud 'clack'.

   a)   Which of the following statements is true about the balls just before a collision? Put a cross in the box (☒) next to your answer.

   ☐ **A**   The balls have the same velocity.

   ☐ **B**   The balls have the same momentum.

   ☐ **C**   The balls have the same kinetic energy.

   ☐ **D**   The balls have the same acceleration.

   (1 mark)

   b) **i)**   The collision of the clacker balls is inelastic. Explain what this means.

   ....

   ....

   (2 marks)

   ii)   Explain how total momentum is conserved before and after the collision.

   ....

   ....

   (2 marks)

   c)   The kinetic energy of one ball just before it collides is 0.024 J. The ball has a mass of 0.012 kg. Calculate the momentum of the ball.

   Momentum ....kg m/s

   (3 marks)

   (Total for Question 1 = 8 marks)

Electrocardiograms

2

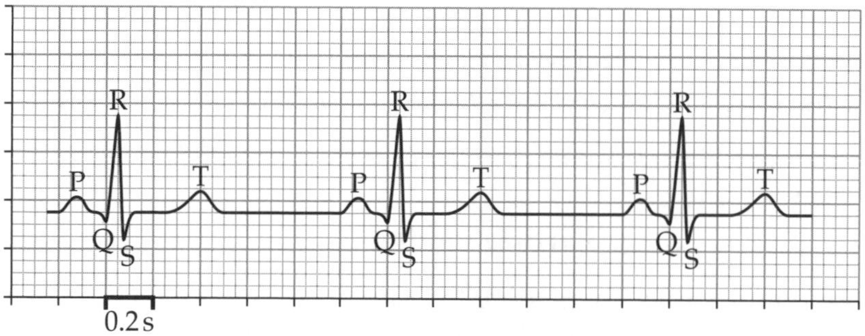

0.2 s

a) The diagram shows an electrocardiogram (ECG). What is measured by an EGC? Put a cross in the box (☒) next to your answer.

☐ **A** action potentials in the heart

☐ **B** blood pressure in the heart

☐ **C** sound of valves opening in the heart

☐ **D** body temperature

(1 mark)

b) Use the graph to calculate the number of beats per minute of the heart.

Heartbeat rate .... beats per minute

(2 marks)

c) Describe two ways that an ECG can give evidence of heart disease.

....

....

(2 marks)

d) Some people are fitted with pacemakers. Describe what a pacemaker does.

....

....

(2 marks)

e) A pulse oximeter fixed to a finger of a patient is a simpler method of monitoring heart rate than an ECG. State the other information that the pulse oximeter gives.

....

(1 mark)

(Total for Question 2 = 8 marks)

Refraction and reflection

3 The diagram shows an endoscope. It is made up of a bundle of optical fibres.

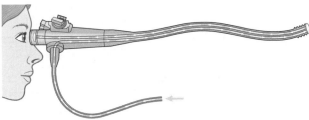

a) i) State a medical use for an endoscope.

....

(1 mark)

ii) When the endoscope is bent, light rays hit the sides of the optical fibres. Explain why the light does not escape from the optical fibres.

....

....

(2 marks)

b) The eyepiece lens at the end of the endoscope produces a magnified image.

i) Explain how light rays are bent as they enter and leave the lens.

....

....

(2 marks)

ii) The focal length of the eyepiece lens is 0.15 m. Calculate the power of the lens.

Power of lens .... dioptres

(2 marks)

iii) The object is 0.08 m from the eyepiece. Calculate the distance of the image from the eyepiece.

Distance .... m

(3 marks)

(Total for Question 3 = 10 marks)

X-rays

4   X-rays are used for imaging in medicine and elsewhere. They can be produced by an X-ray tube.

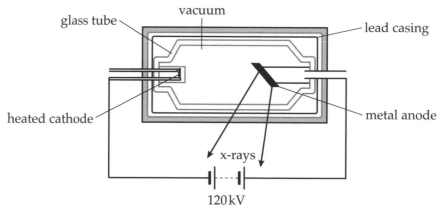

a)   Match up the functions with the parts of the X-ray tube that carry them out as shown in the diagram. Draw lines between the function and the part.

Function   **Part**

| stops air entering the tube |
| thermionic emission of electrons |
| prevents X-rays escaping except through the window |

- heated cathode
- metal anode
- lead lining
- glass tube

(2 marks)

b)   Explain why the tube is evacuated.

....

....

(2 marks)

c)   Explain why there is a high potential difference between the cathode and the anode.

....

....

(2 marks)

d) i)   Calculate the kinetic energy of the electrons as they are about to hit the metal target when there is a voltage of 120 kV between the anode and cathode. The charge on an electron is 1.6×10^{-19} C.

Kinetic energy: ....J
(2 marks)

ii)   Calculate the number of electrons required to hit the target to maintain a current of 6.4×10^{-5} amperes in the circuit.

Number of electrons: ....
(2 marks)

(Total for Question 4 = 10 marks)

Radioactivity in medicine

5   A number of radioactive isotopes are used in medicine. One recently developed technique is called Targeted Alpha Therapy (TAT). In this method an alpha-emitting radioactive isotope is introduced into cancer cells. The alpha radiation kills the cells. One isotope used in TAT is lead-192.

    a) Alpha particles are made up of protons and neutrons. Complete the table below showing the properties of protons and neutrons.

Particle	Proton	Neutron
Relative mass	1	
Relative charge	+1	

          (2 marks)

    b) Describe the effect that alpha decay has on the lead-192 nucleus.

    ....

    ....

          (2 marks)

    c) Complete the equation below for the nuclear reaction that occurs.

$$^{192}_{82}\text{Pb} \quad \rightarrow \quad ^{\square}_{\square}\text{Hg} \quad + \quad ^{4}_{2}\text{He}$$

          (2 marks)

    d) Compare the medical and ethical advantages of using TAT to destroy tumours with other types of radiotherapy using beta ray emitters internally and gamma ray emitters externally.

    ....

    ....

    ....

    ....

    ....

    ....

    ....

    ....

    ....

    ....

    ....

    ....

    ....

    ....

    ....

    ....

          (6 marks)

          (Total for Question 5 = 12 marks)

Pressure and volume of gases

6   A student used the apparatus shown below to investigate how the pressure in a gas changed when its volume was changed.

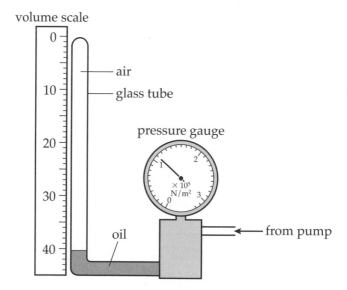

The student used a bicycle pump to increase the pressure and read the pressure gauge when the oil was at particular levels. The table below shows the results that the student collected.

Volume of air (cm³)	Pressure ($\times 10^5$ Pa)	1/pressure
40	1	1
30	1.33	
20	2	0.5
15	2.67	0.374
10	4	0.25

a) When analysing the results the student calculated the value 1/pressure for the readings but left one value out. Complete the table.

(1 mark)

b) The student plotted the data onto the grid shown. Finish plotting the results, draw a line of best fit and extend it until it reaches the axes.

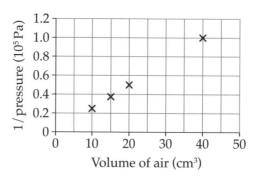

(3 marks)

c) Explain the relationship between volume and pressure of a gas using the kinetic theory model.

....

....

(2 marks)

d) Another student carried out the experiment with similar equipment over a
few days. The graph the second student drew is shown below.

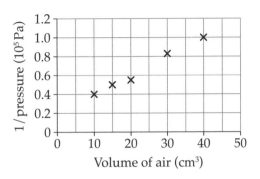

Compare the graph obtained by the second student with the first, giving possible
explanations for any differences.

....

....

....

....

....

....

....

....

....

....

....

....

....

....

....

....

....

(6 marks)

(Total for Question 6 = 12 marks)

Formulae

You may find the following formulae useful:

intensity $= \dfrac{\text{power of incident radiation}}{\text{area}}$ $\qquad$ $I = \dfrac{P}{A}$

power of lens $= \dfrac{1}{\text{focal length}}$

The relationship between focal length, object and image distance $\qquad$ $\dfrac{1}{f} = \dfrac{1}{u} + \dfrac{1}{v}$

current = number of particles per second $\times$ charge on each particle $\qquad$ $I = Nq$

kinetic energy = electronic charge $\times$ accelerating potential difference $\qquad$ $KE = \tfrac{1}{2}mv^2 = e \times V$

frequency $= \dfrac{1}{\text{time period}}$ $\qquad$ $f = \dfrac{1}{T}$

The relationship between temperature and volume for a gas $\qquad$ $V_1 = \dfrac{V_2 T_1}{T_2}$

The relationship between volume and pressure for a gas $\qquad$ $V_1 P_1 = V_2 P_2$

The relationship between the volume, pressure and temperature for a gas $\qquad$ $\dfrac{P_1 V_1}{T_1} = \dfrac{P_2 V_2}{T_2}$

Periodic table

Key

| relative atomic mass |
| **atomic symbol** |
| name |
| atomic (proton) number |

| 1 |
| **H** |
| hydrogen |
| 1 |

1	2											3	4	5	6	7	0
																	4 **He** helium 2
7 **Li** lithium 3	9 **Be** beryllium 4											11 **B** boron 5	12 **C** carbon 6	14 **N** nitrogen 7	16 **O** oxygen 8	19 **F** fluorine 9	20 **Ne** neon 10
23 **Na** sodium 11	24 **Mg** magnesium 12											27 **Al** aluminium 13	28 **Si** silicon 14	29 **P** phosphorus 15	31 **S** sulfur 16	35.5 **Cl** chlorine 17	40 **Ar** argon 18
39 **K** potassium 19	40 **Ca** calcium 20	45 **Sc** scandium 21	48 **Ti** titanium 22	51 **V** vanadium 23	52 **Cr** chromium 24	55 **Mn** manganese 25	56 **Fe** iron 26	59 **Co** cobalt 27	59 **Ni** nickel 28	63.5 **Cu** copper 29	65 **Zn** zinc 30	70 **Ga** gallium 31	73 **Ge** germanium 32	75 **As** arsenic 33	79 **Se** selenium 34	80 **Br** bromine 35	84 **Kr** krypton 36
85 **Rb** rubidium 37	88 **Sr** strontium 38	89 **Y** yttrium 39	91 **Zr** zirconium 40	93 **Nb** niobium 41	96 **Mo** molybdenum 42	[98] **Tc** technetium 43	101 **Ru** ruthenium 44	103 **Rh** rhodium 45	106 **Pd** palladium 46	108 **Ag** silver 47	112 **Cd** cadmium 48	115 **In** indium 49	119 **Sn** tin 50	122 **Sb** antimony 51	128 **Te** tellurium 52	127 **I** iodine 53	131 **Xe** xenon 54
133 **Cs** caesium 55	137 **Ba** barium 56	139 **La*** lanthanum 57	178 **Hf** hafnium 72	181 **Ta** tantalum 73	184 **W** tungsten 74	186 **Re** rhenium 75	190 **Os** osmium 76	192 **Ir** iridium 77	195 **Pt** platinum 78	197 **Au** gold 79	201 **Hg** mercury 80	204 **Tl** thallium 81	207 **Pb** lead 82	209 **Bi** bismuth 83	[209] **Po** polonium 84	[210] **At** astatine 85	[222] **Rn** radon 86
[223] **Mn** francium 87	[226] **Ra** radium 88	[227] **Ac*** actinium 89	[261] **Rf** rutherfordium 104	[262] **Db** dubnium 105	[266] **Sg** seaborgium 106	[264] **Bh** bohrium 107	[277] **Hs** hassium 108	[268] **Mt** meitnerium 109	[271] **Ds** darmstadtium 110	[272] **Rg** roentgenium 111							

Elements with atomic numbers 112–116 have been reported but not fully authenticated

* The lanthanoids (atomic numbers 58–71) and the actinoids (atomic numbers 90–103) have been omitted.

The relevant atomic masses of copper and chlorine have not been rounded to the nearest whole number.

Answers

Biology answers

1. Rhythms

1 a) The minimum number of hours of light needed to make the plant flower is 12 (1) because plants with less light than this did not flower (1).

b) Plants should be genetically identical (1) and there should be the same light intensity each time (1). (*It would also be possible to say that the same temperature, same concentration of carbon dioxide, same level of watering or same level of mineral ions should be used.*)

c) more flowers for bees/pollinators to visit (1); increased chance of pollination (1)

2 a) A daily rhythm (1).

b) Melatonin levels would drop in the morning (1); and stay low during the day (1); so that we feel more awake during the day (1).

c) Because the melatonin is secreted (1); when it would be night where the person flew from (1).

2. Plant defences

1 a) organism that causes disease (1)

b) to act as a comparison/to show that any difference in the two outcomes must be due to the presence of the garlic (1)

c) the control (1)

d) Any two from: a lid was placed over the tubes to stop other bacteria from getting in/to stop contamination with possible pathogens (1); to stop organisms entering that would kill the bacteria (1); to stop escape of bacteria from test tube (1).

e) Any two from: temperature (1); same volume of garlic juice and water (1); same species of bacteria (1); same number of bacteria (1); same amount of jelly in each tube (1).

f) The plant is able to defend itself from attack by pathogens (1).

2 a) Any two from: toxic chemicals (1); bitter tasting chemicals (1); spines on leaves to deter feeding (1).

b) Because they eat leaves (1); reducing surface area (1); for photosynthesis (1).

3. Growing microorganisms

1 a) 20; 40; 80 (1); *As number doubles every 20 minutes then at time 0 there were 10, at 20 min it is 2 × 10 = 20, at 40 min it is 2 × previous answer (20) = 40, at 60 min it is 2 × 40 = 80.*

b) exponential (1)

2 a) milk heated to a high temperature (1); for a short period of time (1)

b) Some bacteria are still alive (1); (the milk) contains food/energy source for bacteria to grow (1).

c) open carton on work top (1); opening allows route for bacteria/microorganisms to enter (1); warmer conditions, which encourage bacteria to grow and reproduce so numbers higher (1).

4. Vaccines

1 a) caught cowpox (1); caught smallpox (1); did not develop smallpox (1)

b) Any two from: virus entered the body/cells (1); cells damaged by virus (1); person not immune/antibodies not present. (1)

c) less likely to catch smallpox (1); the cowpox antigen produced memory lymphocytes (1); exposure to similar antigen/smallpox antigen (1); caused secondary response to occur (1)

2 a) On the surface/cell (surface) membrane (1); of cells/bacteria/other named cell (1).

b) Any three from: to get antigen into blood (1); to stimulate immune response (1); to produce memory lymphocytes (1); to make person immune (1).

5. Antibodies

1 a) antigens on the surface of pathogens (1)

b) another/subsequent/second exposure (1); to the same antigen (1)

c) Any three from: to divide if exposed to same antigen again (1); to produce antibodies (1); specific/complementary to the antigen (1); rapidly (1).

2 a) Antibodies of one type (1); produced in large quanities by hybridoma cells (1).

b) Any four from: organism exposed to a disease/antigens (1); organism produces lymphocytes (1); fused with (1); tumour cells (1); to form hybridoma (that makes monoclonal antibodies) (1).

c) Antibody/hormone is not correct shape to recognise hormone/antibody (1).

6. The kidneys

1 a) A: renal artery (1); B: kidney (1); C: renal vein (1)

b) B: to clean/filter the blood/removing urea (1); C: to transport (cleaned) blood away from kidney (1)

c) ureters (1)

2 break down of excess (1); amino acids (1); in liver. (1)

3 Treatment 1: Kidney dialysis (1); blood removed from body to be cleaned/filtered/urea removed (1); Treatment 2: Kidney transplant (1); new/donor kidney put into body (1).

7. Inside the kidneys

1 Glomerulus and Bowman's capsule – filtration (1); Collecting duct – reabsorption of water (1)

2 a) Urea moves into the glomerulus and then into the Bowman's capsule. Large proteins do not pass into the glomerulus because they are too big to fit through the walls of the capillary.

b) red blood cell – too large to enter Bowman's capsule/leave capillary/leave glomerulus/pass through membrane (1); glucose – it is reabsorbed (1)

3 to reabsorb back into the blood/capillary (1); glucose for respiration/molecules useful to the body (1)

8. The role of ADH

1 a) The volume of urine would be smaller (1); the urine would become more concentrated (1).

b) Any four from: water lost from body as sweat (due to running/hot day) (1); (therefore) water content of blood decreased (1); pituitary (gland) responded by secreting more ADH (1); so more in the blood stream (1); so that more water can be reabsorbed from kidney/nephron/collecting duct back into the blood (1).

2 a) pituitary (gland) (1); blood (stream) (1); kidney/nephron/collecting duct (1)

b) pituitary = site of release (1); blood = idea of being in transit (1); kidney/nephron/collecting duct = site where it binds/target organ/causes water to be reabsorbed (1)

9. The menstrual cycle

1 a) 28 (1); menstruation/period (1); repair/build up (1); 14 (1); egg (1); ovary (1)

b) FSH stimulates/causes follicles to grow and mature (1); LH causes ovulation (1).

c) oestrogen (1); progesterone (1)

2 a) menstruation (1); ovulation (1)

b) Any time between day 14 and about day 17 (1).

c) from the diagram can see that lining of uterus remains in place beyond 28 days (1); would expect lining of uterus to start to break down {at about day 28/before day 35) (1)

10. Hormone control

1 a) oestrogen (1); LH (1); progesterone (1)

b) stimulates follicles to mature (1); stimulates ovaries to release oestrogen (1)

c) Lining of uterus would not be repaired (1); LH levels will not increase/LH will not be released (1); ovulation would not occur/person would be infertile (1).

2 Any four from: negative feedback means that a change in one factor will lead to the opposite change (1); for example during the menstrual cycle high levels of oestrogen cause release of LH (1); the surge in LH levels causes ovulation (1); the corpus luteum is left behind and starts to produce progesterone (1); progesterone inhibits the release of LH and levels fall again (1).

11. Fertilisation

1 that a couple are unable to have children (1)

2 a)

Description of infertility treatment	Name of infertility treatment
Joining of a sperm and an egg in a dish	*In vitro* (1) fertilisation (1)
Chemicals given to a woman to encourage ovaries to release eggs	Hormones (1)
Name of woman into whom a genetically unrelated embryo is placed	Surrogate (1)

b) Advantage = childless couple can have a child (1) Disadvantage = may be difficult for surrogate woman to give up baby/cost of surrogacy (1)

3 a) Two marks from similarities: contain cytoplasm (1); contain a haploid nucleus/23 chromosomes (1); presence of mitochondria (1); presence of a cell membrane (1). Three marks from differences: only the sperm has a tail (1); only a sperm has an acrosome (1); mitochondria are found packed into the middle section (1); only the egg has a large nutrient store (1); egg has more cytoplasm (1).

b) Any two from: sperm tail = to move sperm towards egg/ to swim (1); acrosome = to contain enzymes/to digest outer layers of egg (1); mitochondria = to supply energy for tail/ swimming (1); few mitochondria = so sperm light (1); large nutrient store = for cell division/mitosis (1); egg has more cytoplasm = to store nutrients/for many organelles (1).

4 Any four from: one woman supplies eggs (1); which are fertilised by sperm from infertile couple (1); using IVF (1); fertilised eggs allowed to develop into embryos (1); one or two healthy embryos are placed in uterus of the infertile woman (1).

12. Sex determination

1 a) X (1); the girl has two X chromosomes, one from each parent. (1)
   b) i) 1 mark for parental sex chromosomes and 1 mark for all possible childrens's chromosomes

Father

		X	Y
Mother	X	XX	XY
	X	XX	XY

   ii) female (1)

2 a) 50%/½/0.5 (1); depends on which sperm fertilises the egg (1); as half the sperm will carry a male sex chromosome/Y chromosome and half the sperm will carry a female sex chromosome/X chromosome (1).
   b) The statement is not correct (1); the probability of having a child who is a boy is always 50% (1).

13. Sex-linked inheritance

1 The gene for haemophilia is carried on the X chromosome (1); but is absent from the Y chromosome (1)

2 a) Genetic diagram to show: Male genotype as X^hY (1); female genotype as X^HX^h or X^HX^h (1); gamete for each parent (1); possible offspring (1); colour blind female identified/X^hX^h identified (1); Examples of genetic diagrams are:

Father

		X^h	Y
Mother	X^h	X^hX^h	X^hY
	X^h	X^hX^h	X^hY

Father

		X^h	Y
Mother	X^H	X^HX^h	X^HY
	X^h	X^hX^h	X^HY

Shaded cells show colour blind females.
   b) only one X^h allele is required in males to show condition (1); chance lower/less likely to have two X^hX^h alleles in female combining which is needed to show condition (1)

14 and 15. Biology extended writing 1 and 2

Answers can be found on page 129.

16. Courtship

1 mate for life = swans; several mates over one season = male lion; one mate for a breeding season = emperor penguins (3 marks for all 3 correct, 2 mark if only 2 correct and 1 or 1 correct.)

2

Example	Gender displaying	What is the selection being based on?
Lion	male	mane
Peacock	male	tail

(1 mark for all genders correct and 1 mark for each of the features)

3 The male that can dance well is likely to be fit and healthy (1); this means that offspring are likely to be fit and healthy as well (1).

17. Parenting

1 a) 50:50/1:1/50%/half and half (1)
   b) Builds a nest (1); to protect the offspring (1).

2 a)

Description of poison dart frog parental care behaviour	Parental care behaviour example
Frog uses legs to spread water over eggs to keep them damp	protecting the young from harm (1)
After hatching, a parent carries the tadpoles to different places to separate them to stop the tadpoles eating each other	protecting the young from harm (1)
The female visits the tadpoles and lays unfertilized eggs for them to eat	helping the young to find food (1)

b) They increase the chance of offspring surviving (1); (and when mature) passing genes onto the next generation. (1)

3 a) Increased chance of survival of offspring/example quoted such as supplying food/warmth/protection (1); increased chance of parental genes being passed on by the offspring. (1)
   b) They are a long way from food/sea/increased likelihood of starvation (1); could die due to very cold temperatures. (1)

18. Simple behaviours

1 a)

Description of animal behaviour	Example is an innate behaviour or imprinting
Baby able to suckle for milk very soon after birth	innate (1)
Young goose bonding with the first moving object it sees	imprinting (1)
A chick using its 'tooth' on the end of its beak to crack open the shell so it can hatch out	innate (1)
A butterfly being able to fly after emerging from its cocoon/chrysalis	innate (1)

   b) automatic behaviour/one that the animal does not have to learn (1)

2 Lorenz = imprint (1); Tinbergen = innate (1)

3 a) 0 and 10 (1). (*The answer should be rounded to the nearest whole number.*)
   b) Rapid movement increased chance of woodlice entering damp/leaving dry half (1); hence all 5 moved out of this half/moved into the damp half (1); slower movement in damp half increased the time spent in this half (1); hence all 5 stayed in this half (1).

19. Learned behaviour

1 a) A (1); D (1)
   b) It is a learned behaviour (1); in which the response is 'switched off' (1); due to a repeated stimulus. (1)

2 Innate response shown due to normal stimulus = Salivating when presented with food (1); Additional stimulus supplied = Bell rung with food presentation (1); An association of additional stimulus with normal stimulus = Bell means food presentation (1); Conditioned behaviour shown = Salivating when bell rung (1)

3 Any four from: gill not withdrawn due to buffeting by waves/ seaweed/other named example (1); gill not withdrawn means more oxygen can be extracted for sea slug (1); for respiration/ energy release (1); saves energy (used for withdrawing gills) (1); sea slug can concentrate on potential dangers (1); sea slug can deal with unexpected changes in the environment (1).

20. Animal communication

1 a) sound (1); chemical (1); visual (1)
   b) for chemical = pheromones in moths (1); to attract a mate/ influence behaviour of other individuals (1)
   for visual = two from: flashing lights in glow-worms (1); to attract the attention of males (1); facial expression/body language in mammals/named mammal such as a cat (1); to show mood (1); gestures/gesture described such as waving (1); to show presence (1)

2 a) Goodall = chimpanzees (1); Fossey = gorillas (1)
   b) Any three from: observed and recorded behaviours (1); of primates (1); showed that primates have complex social interactions/groups (1); primates have many different calls (for communication). (1)

3 Bird song to declare territory/courtship (1); stag roaring to advertise his strength (1); grasshoppers making a high-pitched sound to attract a mate (1). (You would get marks for other examples but you must give both sound signal and what is being communicated for each mark.)

21. Plant communication

1 a) The insects are most likely to come from the south (1); because wind carries the scent from that direction (1).
   b) releasing chemicals into the air (1) when being attacked by herbivores to warn surrounding plants (1) (If you give other correct examples in an exam, you would get the marks.)
   c) Any two from: bright colours (1); mimic shape of pollinator species (1); guiding lines on petals (1).

2 Any three from: idea that co-evolution is a change in characteristics/evolution of one species due to/caused by a change in another species (1); examples include: long tongue linked to access nectar in long flower (1); position of pollen enables pollen to be deposited on hummingbird (1); nectar rich in sugars to supply energy for hummingbird hovering/fast wing movement (1); plants bloom/energy-rich nectar presence to coincide with energy-demanding hummingbird breeding season (1).

22. Human evolution

1 Any three from: toe arrangement (1); length of arms (1); height (1); way they walk (1).

2 a) Any two from: negative correlation/as years before present became less, brain volume increased OR positive correlation – as time 'increases' brain volume increases (1); greatest increase in brain volume between 2.4 and 1.8 million years ago (1); increase in brain volume not linear, increased by $500\,cm^3$ in 2.6 million years (1).
   b) Pour sand into brain case until full (1); pour this sand into a measuring device/measuring cylinder to work out volume. (1)
   c) increased brain volume/size (1); to at least $550\,cm^3$ (1)

3 a) A (1); because B has more chips (1); so more advanced/more complex construction (1)
   b) Any three from: Smooth area in palm of hand (1); will not cut/damage hand (1); chipped section away from hand (1); (as it) has sharp edges (1); for cutting/unlike smooth area. (1)

23. Human migration

1 When the climate cooled down in the ice age sea levels dropped/fell (1); this meant that there was less water and crossing out of Africa was easier (1).

2 a) in mitochondria (1); which are in the cytoplasm (1)
   b) Mitochondrial DNA is less likely to have degraded over time (1); mitochondrial DNA is more abundant in cytoplasm (1).
   c) Mitchondria in cytoplasm/eggs (1); sperm do not donate mitochondria (1); female supplies the egg (1).
   d) The African Eve theory suggests that there was a common ancestor for all people who originated in Africa (1); evidence for this theory comes from mtDNA from people all over the world (1); who all show the same DNA in an unbroken line (1).
   e) Zero (1); Y chromosome delivered by sperm (1); which does not donate any mitochondria/mitochondrial DNA. (1)

24. Biology extended writing 3

Answers can be found on page 129.

25. Biotechnology

1 a) A = culture broth (1); B = sensors (1); C = stirrer (1)
   b) to remove excess heat/keep temperature constant (1); to maintain optimal temperature for microorganism growth (1)
   c) supplies oxygen (1); for aerobic respiration (1); not contaminated (1)

2 a) i) steam forced in (1); kills microorganisms (1)
      ii) Two from: Other microorganisms could compete with useful microorganism (reducing product formed) (1); other microorganisms could contaminate product (so not useable) (1); other microorganisms could be pathogenic/harmful (1).
   b) Any three from: stirrer would stop moving (1); uneven distribution of nutrients/non-optimal conditions in much of fermenter (1); microorganisms sink to bottom (1); less product/penicillin formed (1).

26. Microorganisms for food

1 a)

Name of food produced by microorganism	Type of microorganism
bread	yeast (1)
mycoprotein	fungus (1)
wine	yeast (1)
yoghurt	bacterium (1)

   b) *Fusarium* (1)

2 a) to supply the bacteria (1)
   b) Any three from: so no other (living) microorganisms present (1); to compete (with yoghurt-making bacteria) (1); to contaminate product/yoghurt (1); that could cause harm (1)
   c) i) thickens (1); as milk protein (1); solidifies/coagulates (1)
      ii) Any five from: as time increases, the pH decreases (1); no change in pH for the first 30 min (1); greatest decrease in pH between 90–120 min (1); bacteria in yoghurt carrying out anaerobic respiration/fermentation (1); producing lactic acid (1); which lowers the pH. (1)

27. Mycoprotein

1 a) $(6/16) \times 100 = 37.5\%$ (1)
   b) There are 76.0 milligrams/thousandths of a gram (1); of cholesterol (1); in (every) 100 g of minced beef. (1)
   c) If we take in too much energy we can gain weight (1); this can lead to health complications such as heart disease/Type 2 diabetes.

2 a) Rapid population growth (1); production independent of climate (1); use of waste products from other industrial processes (1).
   b) Any two from: to supply nitrogen (1); for amino acids (1); for growth (1).
   c) to supply oxygen (1); for aerobic respiration (1)

28. Enzyme technology

1 Any two from: easier to separate enzyme from product (1); so enzyme can be reused (1); more temperature stable so can use a higher temperature (1); so reaction proceeds more rapidly (1).

2 Invertase catalyses the breakdown of sucrose (1) into glucose and fructose (1).

3 a) chymosin gene (1); placed into bacteria (or yeast) (1); bacteria make chymosin/protein (1)
   b) Two from: converts milk protein (1); into curds and whey (1); to make cheese (1).
   c) (cheese) suitable for vegetarians (1)

4 Enzymes digest food stains on the clothes (1); and they work efficiently at low temperatures (1).

29. DNA technology

1 a) to supply optimum growth/reproduction conditions/named example described (1)
   b) The restriction enzymes cut out the human gene (1) for insulin and leave sticky ends (1) which means the gene can splice together with the bacterial DNA (1).
   c) Any two from: to produce matching sticky ends (1); complementary bases (1); so insulin gene can be incorporated into plasmid (1).
   d) circle of DNA (1); with insulin/added gene in it (1)
   e) Any three from: added nutrients not wasted (or example described) (1); no competition (1); no contamination (1); from bacteria without modified plasmids/other microorganisms (1).

30. Global food security

1 a)

Pest management strategy	How it works
Using pheromone traps	Pheromone attracts pests to the trap (1)
Attracting natural predators	Predators attack/kill pests (1)

   b) Any two from: more successful (1); reduced chance of pesticide resistance forming (1); less damaging to the environment (1).

2 a) Advantages: biofuels are renewable/fossil fuels are not renewable (1); biofuels remove carbon dioxide from the air whilst growing (1).
   b) Disadvantage: may reduce food production (1); as growing the biofuel plants uses land that could be used for food (1).

3 a) high yield so can feed more people (1); low fertiliser requirement so no need to apply fertiliser/reduce cost (1); pest resistant (or example given) so less pest damage/do not need to apply pesticide (1)
   b) drought resistant to cope with times of water shortage without dying (1); tolerant of high temperature so can be grown in hot areas of Africa (1)
   c) less likely to be blown over in the wind/less likely to snap/plant uses less energy in growing the stem, so has more to use for making seeds (1)

31. A GM future?

1 a) plant containing genetic material/DNA/gene (1); from another species (1)
   b) microorganism/*Agrobacterium* sp. (1); that transfers gene (from one species) (1); to another species (1)

2 a) Some evidence from experiments show that mice with cancer live longer if given flavonoids (1); (therefore people with cancer) may want to eat these tomatoes because the flavonoids may be good for them/help them fight cancer (1).
   b) One from: has not been tested on humans (1); it may not work for all cancers (1); people may not be able to afford them (1).
   c) Any five from: tomato plant cells/callus/disc infected with this *Agrobacterium* (1); *Agrobacterium* is acting as a vector (1); plasmid (with flavonoid gene) inserted into plant DNA (1); plant grows/forms (from callus) (1); with its cells containing flavonoid gene (1); gene active/gene makes flavonoids in fruit/tomato (1).

32. Insect-resistant plants

1 a) (chemical) that is poisonous to/kills (1); to insects/named insect, e.g. caterpillar (1); to reduce damage to plants (1)

b) (One type of bacterium) contains/supplies the Bt gene (1); (a different bacterium) has Bt gene added to its plasmid (1); acts as a vector/adds the gene to plant cells (1).

2 a) increased yield/less damage by insects/pests (1); less chemical insecticide used which could be harmful (1).
  b) i) Pollen transferred from insect-resistant/transgenic plant (1); to closely related wild plants (1).
     ii) Two from: insects associated with the wild plants could be killed (1); (so) fewer pollinators (available) (1); increase chance of insects/pest becoming resistant to the Bt toxin (1).

33 and 34. Biology extended writing 4 and 5
Answers can be found on page 129.

Chemistry answers

35. Water testing
1 sodium − yellow (1); potassium − lilac (1); copper (II) − green/blue (1)
2 a) A qualitative test identifies if something is present. (1)
  b) red (1) (Brick red and orange-red would also be accepted.)
  c) white precipitate (1); does not dissolve when more sodium hydroxide added (1)
3 a) i) iron (II) chloride − green precipitate (1)
     ii) iron (III) chloride − red-brown precipitate (1)
  b) $FeCl_2$ (aq) + 2NaOH (aq) → $Fe(OH)_2$ (s) + 2NaCl (aq) (formulae, 1; balancing, 1; state symbols, 1)

36. Safe water
1 a) silver nitrate (1)
  b) He needs to add nitric acid (1); this acid reacts with any carbonate ions present, as these ions also give a precipitate with silver nitrate (1).
  c) cream precipitate (1)
2 a) NH_4^+ (1)
  b) sodium hydroxide (1)
  c) ammonia (1); test with damp red litmus (1); which turns blue (1)
3 a) silver chloride (1)
  b) Ag^+ (aq) + Cl^- (aq) → AgCl (s) (formulae, 1; state symbols, 1)

37. Safe limits
1 dissolved (1); lilac (1); precipitate (1)
2 a) flame (1)
  b) i) I would bubble the gas through a solution of limewater (1); the solution would turn milky (1).
     ii) quantitative (1); how much of the carbonate ion is present (1)
  c) correct dose/might be toxic (1)
3 a) They form white precipitates (1).
  b) add more sodium hydroxide (1); aluminium hydroxide precipitate dissolves (1)
  c) The solution contains water and calcium chloride (1); so is a mixture of the two substances, and not pure (1).

38. Water solutes
1 ticks in calcium and in magnesium boxes; crosses in other three boxes (4)
2 a) scum/calcium stearate (1)
  b) Any two from: calcium/magnesium ions in water (1); react with soap (1); to form an insoluble substance (1).
  c) no lather (1); until lots of soap added (1)
  d) 5 g/2 dm³ (1) = 2.5 g dm⁻³ (1)
3 a) 250 cm³ = 250/1000 dm³ = 0.25 dm³ (1); mass of substance = 0.25 dm³ × 25 g dm⁻³ = 6.25 g (1)
  b) new volume = 1000 cm³ or 1 dm³ (1); and new concentration is therefore 6.25 g dm⁻³ (1) (as there is still the same mass of sodium chloride present)

39. Hard and soft water
1 a) i) boiling (1)
     ii) limescale (1)
  b) calcium sulfate (1)
2 a) prevent formation of limescale (1); which blocks pipes/boilers (1) OR hard water wastes soap (1); so softening uses less soap (1) (Note that both marks must come from the same explanation.)
  b) The resin contains sodium ions (1); held on beads (1); which swap places for calcium ions (1).
  c) rinse through with strong sodium chloride solution (1)
3 Calcium hydrogencarbonate in temporary hard water (1); decomposes on heating (1); to produce limescale/calcium carbonate (1); equation: $Ca(HCO_3)_2 → CaCO_3 + CO_2 + H_2O$. (1)

40. Moles and mass
1 a) B (1)
  b) B (1)
2 a) (1 × 12) + (2 × 16) = 44 (1)
  b) i) 88/44 = 2 (1)
     ii) 88 kg = 88 000 g (1); 88 000/44 = 2000 (1)

3 RFM of calcium carbonate = 40 + 12 + (3 × 16) = 100; RFM of calcium oxide = 40 + 16 = 56 (1); moles of calcium carbonate = 84/100; calcium oxide = 84/56 (1); so, more moles of CaO (1). *It is possible to answer this question without the calculation:* the relative formula mass of calcium carbonate is larger than the relative formula mass of calcium oxide (1); because it has an extra carbon and two oxygen atoms (1); and as there is the same mass of both, there must be more moles of CaO (1).

4 A hydrogen atom has a relative atomic mass of 1 (1); hydrogen molecules have a relative formula mass of 2 (1); so 2 g of gas contains 2 moles of atoms and 1 mole of molecules (1).

41. Moles in solution
1 sodium hydroxide: 23 + 1 + 16 = 40 (1); hydrogen chloride: 1 + 35.5 = 36.5 (1); sulfur dioxide: 32 + (2 × 16) = 64 (1)
2 a) 1.85 g ÷ 1 dm³ = 1.85 g/dm³ (1)
  b) RFM of calcium hydroxide = 40 + (2 × 16) + (2 × 1) = 74 (1); number of moles = 1.85 g/74 = 0.025 (1)
  c) 0.025 mol ÷ 1 dm³ = 0.025 mol dm⁻³ (1)
3 a) sodium hydroxide has RFM of 23 + 1 + 16 = 40 (1); first solution: 20/40 moles in 500/1000 dm³; so 0.5 moles ÷ 0.5 dm³ = 1 mol/dm³ (1); second solution: 32/40 moles in 750/1000 dm³; so 0.8 moles ÷ 0.75 dm³ = 1.07 mol/dm³ (1); second solution is more concentrate (1)
  b) number of moles = volume in cm³ ÷ 1000 × concentration in mol/dm³; first solution: 200 cm³ ÷ 1000 × 0.5 mol/dm³ = 0.1 moles (1); second solution: 500 cm³ ÷ 1000 × 0.25 mol/dm³ = 0.125(1); second has more HCl (1)

42. Preparing soluble salts 1
1 sulfuric − sulfate (1); nitric − nitrate (1); hydrochloric − chloride (1)
2 a) carbon dioxide (1)
  b) all the sulfuric acid used up (1); so the reaction with nickel carbonate stops (1)
  c) filtration (1)
  d) nickel carbonate + sulfuric acid → nickel sulfate + carbon dioxide + water (name of salt, 1; remainder of equation, 1)
3 a) $Ag_2O + 2HNO_3 → 2AgNO_3 + H_2O$ (formula of silver nitrate, 1; other formulae, 1; balance, 1)
  b) evaporate most of the water in the salt solution off (1); then leave concentrated solution in a warm place to crystallise (1)
  c) It is insoluble/only the outer surface of the silver oxide would react (1).

43. Preparing soluble salts 2
1
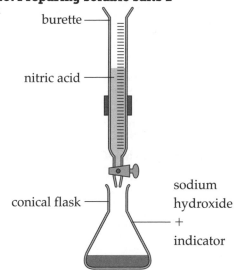

burette

nitric acid

conical flask

sodium hydroxide + indicator

burette (1); conical flask (1); sodium hydroxide (1); indicator (1)
2 a) pipette (1)
  b) Any two from: rinse before use (1); use pipette filler (1); fill to line on stem (1).
  c) HCl + KOH → KCl + H_2O (1 for formula of KCl; 1 for the rest of the equation correct)
3 a) use an indicator (a named indicator, such as litmus or methyl orange would also score the mark) (1); changes colour (if a named indicator is used, then a correct colour change scores the mark too) (1)
  b) H^+ (aq) + OH^- (aq) → H_2O (1)
  c) repeat the experiment (1); until the results are concordant/close to each other (1); and then calculate an average volume (1).

44. Titration calculations
1 a) $NaOH + HNO_3$ (1) → $NaNO_3 + H_2O$ (1)
  b) 28.7 at end (1); 2.20 at start (1); 26.50 added (1)

2 a) $(26.70 + 26.50)/2 = 26.60$ (1)
　b) $(26.70 + 26.50)/2 = 26.60$ (1) (You would also get a mark for calculating an average correctly from wrong answers to part **a)**.)
　c) i) number of moles = volume in cm^3 ÷ 1000 × concentration in mol/dm^3 = 25 cm^3 ÷ 1000 × 0.50 mol/dm^3 (1) = 0.0125 moles (1)
　　ii) the same as **i)**, as the ratio of LiOH to HCl is 1:1 in the equation, so 0.0125 moles (1)
　　iii) 0.0125 moles in (27.5/1000) dm^3, so 0.0125 ÷ 0.0275 (1) = 0.45 mol/dm^3 (1)

45. More calculations from equations

1 a) $25/1000 \times 0.40 = 0.01$ moles (1)
　b) $20/1000 \times 0.50 = 0.01$ moles (1)
　c) $50/1000 \times 0.15 = 0.0075$ moles (1)
2 a) number of moles = 21.5 cm^3/1000 × 0.0015 mol dm^{-3} (1) = 0.0003225 moles (1) (*This answer may be rounded to 0.000323, or given in standard form as 3.225×10^{-4}.*)
　b) the same as in **a)** i.e. 0.0003225 (1); because the calcium ions and carbonate ions react in a 1 to 1 ratio in the chemical equation (1)
　c) 0.0003225 moles in 25/1000 dm^3, so 0.0003225/0.025 (1) = 0.0129 mol dm^{-3} (1)
3 moles KOH = 25 cm^3/1000 × 0.2 mol dm^{-3} = 0.005 moles (1); number of moles of sulfuric acid is half this = 0.0025 moles (2KOH react with H$_2$SO$_4$ in the equation) (1); so, concentration of sulfuric acid = 0.0025 moles × 1000/28.4 cm^3 = 0.088 mol dm^{-3} (1)

46 and 47. Chemistry extended writing 1 and 2
Answers can be found on page 129.

48. Electrolysis

1 a) in solid, ions cannot move (1); water dissolves the solid and allows the ions to move (1)
　b) The streak is caused by the manganate ion (1); which is negatively charged; so moves towards the positive electrode (1)
2 a) reduction − gain of electrons (1)
　b) $2Cl^- \rightarrow Cl_2 + 2e$ (formulae, 1; balancing, 1)
　c) same volumes (1); because to form one mole of gas (1); it takes the same number/2 electrons (1) OR less chlorine (1); it is more soluble than hydrogen (1); so dissolves into water (1)

49. Making and using sodium

1 a) $2Cl^- \rightarrow Cl_2 + 2e$ (1)
　b) They would react together (1); this would make them re-form sodium chloride (1).
　c) liquid (1); has reached its melting point, but not its boiling point (1)
　d) street lamps/coolant in nuclear reactors (1)
2 a) substance that contains ions that are free to move about in electrolysis (1)
　b) $K^+ + e \rightarrow K$ (formula of potassium ion, 1; correct equation, 1)
　c) potassium is very reactive (1); but won't react with argon, as it is a noble gas (1)

50. Electrolysis of salt water

1 brine (1); chloride (1); cations (1)
2 row 1: copper, chlorine (1); row 2: copper, oxygen (1); row 3: hydrogen, bromine (1); row 4: hydrogen, oxygen (1)
3 a) fizzing/bubbles of gas (1)
　b) Any two from: hydrogen ions (1); from water molecules (1); discharged in preference to sodium ions (1).
　c) $2H^+ + 2e \rightarrow H_2$ (formulae, 1; balance, 1)
　d) removal of hydrogen ions from water leaves hydroxide ions (1); which are alkaline (1)

51. More electrolysis

1

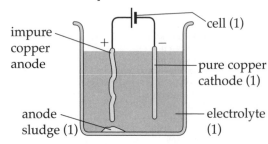

impure copper anode (1)
cell (1)
pure copper cathode (1)
anode sludge (1)
electrolyte (1)

2 a) positive electrode − pure gold (1); negative electrode − item to be plated/copper item (1)
　b) i) $Au^{3+} + 3e \rightarrow Au$ (1)
　　ii) because electrons are gained

3 a) solid copper produced there (1); from copper ions (1); $Cu^{2+} + 2e \rightarrow Cu$ (1)
　b) negative electrode gained $10.63 - 10.25 = 0.38$ g (1); so $9.78 - 0.38 = 9.40$ g (1)

52. Gas calculations 1

1 24 dm^3 × 0.5 = 12 dm^3 (1); 1200 cm^3 /1000 = 1.2 dm^3 and 1.2 dm^3/24 dm^3 = 0.05 moles (1)
2 a) 24 dm^3 × 0.5 = 12 dm^3 (1)
　b) 4800 cm^3 = 4.8 dm^3; number of moles = 4.8/24 = 0.2 moles (1)
3 a) 6×10^{23} (1)
　b) twice as many/$1.2 \times 10^{24}/12 \times 10^{23}$ (1); because each chlorine molecule contains 2 atoms (1)
4 a) 960 dm^3/24 dm^3 = 40 (1)
　b) 2:1 ratio/half as much oxygen (1); so 480 dm^3 (1)
　c) 2 volumes + 1 volume → 2 volumes (1); so 100 dm^3 (1)

53. Gas calculations 2

1 a) RFM iron oxide = 160 (1); 1600/160 = 10 moles (1)
　b) One mole of iron (III) oxide produces three moles of carbon dioxide, so three times as many, i.e. 30 moles (1)
　c) 30×24 dm^3 (1) = 720 dm^3 (1)
2 moles gas = 100/24 000 = 0.00417 (1); the number of moles of Mg in the equation is the same (1); max mass Mg = 100/24 000 (or 0.00417) × 24 (1) = 0.1 g (1)

54. Fertilisers

1 plants (1); manure (1); nitrogen (1); ammonia (1)
2 a) nitrogen − air (1); hydrogen − methane/water (1)
　b) $N_2 + 3H_2 \rightleftharpoons 2NH_3$ (symbols, 1; balancing 1)
　c) reversible (1); so reaction can go both ways (1)
3 a) $NH_3 + HNO_3 \rightarrow NH_4NO_3$ (formulae of starting materials, 1; equation, 1)
　b) They increase plant growth (1); to improve crop yield/grow more crops in the land available (1).
　c) Any four from: washes into ground water (1); encourages growth of plants/algae in rivers (1); eutrophication (1); when plants die, microorganisms break them down, and use oxygen to respire (1); lack of oxygen in water to support other organisms (1).

55. Equilibrium

1 top: tick in right-hand box (1); middle: tick in right-hand box (1); bottom: tick in left-hand box (1)
2 a) dynamic because it is going in both directions (1); and an equilibrium because it is going at the same rate/no change in amounts of substances (1)
　b) increase in temperature favours endothermic/forward reaction (1); so yield is higher (1)
　c) same number of moles on each side (1); so neither side favoured (1)
3 a) moves to right (1); fewer gas molecules that side (1)
　b) paler, so must be more N_2O_4 (1); so low temperature favours forward reaction (1); so forward reaction must be exothermic (1)

56. The Haber process

1 The temperature used is around 450°C (1); the pressure used is around 200 atm (1); and the catalyst used is iron (1).
2 a) increase (1)
　b) no effect (1); just reaches equilibrium position faster/speeds up forward and back reactions by the same amount (1)
3 a) high pressure/1000 atm (1); low temperature/200°C (1)
　b) high pressure very expensive/breaks containers (1); so compromise at 200 atm (1); low temperature gives low rate (1); so need to make hotter around 450°C (1)
　c) catalyst increases the rate (1); as the conditions chosen (e.g. low temperature) make the rate slower than wanted (1)

57. Chemistry extended writing 3
Answers can be found on page 129.

58. Fermentation and alcohol

1 a) D (1)
　b) A (1)
　c) B (1)
2 a) with no oxygen present (1)
　b) needs to be warm enough to make reaction happen (1); but too hot and enzymes (1)
3 a) beer: $550 \times 3.5/100$ (1) = 19.25 ml (1); wine: $150 \times 12.5/100$ = 18.75 ml (1)
　b) $C_6H_{12}O_6 \rightarrow 2C_2H_5OH + 2CO_2$ (formulae, 1; balance, 1)
　c) fractional (1); distillation (1)

59. Ethanol production

1 water (1); H_2O (1); ethene (1); C_2H_4 (1); ethanol (1); C_2H_5OH (1)

2 **a)** crude oil (1)
  b) Any four from: fractional distillation (1); to separate out short and long chain molecules (1); cracking (1); using catalyst (1); of long chain molecules (1); using catalyst (1).
  c) dehydration (1)
  d) high/above 100°C (1); because water is a gas (1)
3 Any four from: oil expensive (1); but can grow sugar cane (1); to ferment (1); cheap/renewable source of ethanol (1); would need to concentrate/purify (1); using fractional distillation (1).

60. Homologous series
1 methane – CH_4 (1); propene – C_3H_6 (1); ethane – C_2H_6 (1)
2 **a)** 4 carbon atoms in a line (1); with 10 hydrogens attached correctly (1)

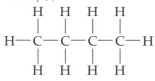

  b) Any three from: series of molecules with same general formula (1); similar chemical properties (1); gradual variation in physical properties (1); differ by CH_2 from each other. (1)
3 **a)** $C_nH_{2n+1}OH$ (keeping OH separate, 1; rest of formula, 1)
  b) C_3H_7Cl (1)
  c) $CH_3OH + 2O_2 \rightarrow CO_2 + 2H_2O$ (formulae, 1; balance, 1)

61. Ethanoic acid
1 **a)** magnesium ethanoate (1); hydrogen (1)
  b) zinc ethanoate (1); carbon dioxide (1)
2 **a)**

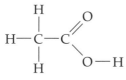

  -COOH drawn out (1); rest of molecule (1)
  b) oxidation (1); of ethanol (1)
3 **a)** C_3H_7COOH (1)
  b) $141 - 118 = 23°C$ (1); $141 + 23 = 164°C$ (1)
  c) $2C_2H_5COOH + Na_2CO_3 \rightarrow 2C_2H_5COONa + CO_2 + H_2O$ (formulae, 1; balance, 1)

62. Esters
1 **a)** A (1)
  b) D (1)
  c) A (1)
2 as a flavouring/aroma (1); in foods/drinks (1)
3 **a)** ethyl propanoate (1)
  b)

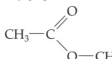

  ester link (-COO-) (1); rest of molecule (1)
  c) lower yield (1); high temperature favours endothermic/back reaction (1)
  d) similarity: produces water/uses an acid (1); difference: product not ionic/alcohol not an alkali (1)

63. Fats, oils and soap
1 solid (1); esters (1); alkali (1); soap (1)
2 **a)** nickel/catalyst (1)
  b) The reaction is called hydrogenation (1); which means that the C=C bonds break and hydrogen atoms are added (1).
  c) increases (1)
3 **a)** hydrophobic = water hating/long carbon chain (1); hydrophilic = water loving/ionic end (1)
  b) heated (1); with strong alkali/sodium hydroxide (1)
  c) hydrophobic ends dissolve in grease (1); hydrophilic ends point out into water (1); and dissolve, lifting grease (1)

64 and 65. Chemistry extended writing 4 and 5
Answers can be found on page 129.

Physics answers

66. Radiation in medicine
1 high-frequency sound – ultrasound scans (1); X-rays – CAT scans (1); visible light – endoscopes (1); gamma rays – PET scans (1)

2 **a)** They all carry energy (1) from a source. (1)
  b) **i)** alpha radiation (1)
      ii) it is made up of particles, the others are all waves (1)
  c) alpha radiation (1); and X-rays (1)
3 The intensity (1) of the sound waves decreases (1) further from the loudspeakers (and becomes less damaging).
4 intensity = $2000\,W/1.6\,m^2$ (1) = 1250 (1) W/m^2 (1) (1.25 kW/m^2 is worth 3 marks with correct working)

67. How eyes work
1 **a)** A cornea, B pupil, C lens, D iris, E ciliary muscles, F retina (all correct, 3 marks; 4 or 5 correct, 2 marks; 2 or 3 correct, 1 mark)
  b) cornea and lens (1) (both needed for one mark, no marks if any other parts are given)
2 **a)** C (1)
  b) There is no limit/greatest distance (1) of an object that the eye can produce a sharp image of/focus on (1).
3 **a)** To form an image the light rays must converge (1) onto the retina (1). The refracting (1) is done by the cornea and lens (1).
  b) (The ciliary muscles contract) making the lens fatter (1); so increasing the refraction/causing the light to bend or converge more (1).

68. Sight problems
1 long (1), short (1), outside (1), fat (1), much (1)
2 **a)** short sight (1)
  b) The lens makes the light rays diverge (1), so that the image moves away from the lens/closer to the retina (1)
  c) **i)** must be clean/free of microbes to avoid infections (1)
      ii) The cells in the cornea need oxygen for respiration/to stay alive (1).
3 **a)** The laser can make very fine/precise cuts/incisions (1) in the cornea; the cornea can be reshaped to form images correctly (1).
  b) One point from each – advantage: correction is permanent; do not have to wear spectacles/contact lenses again (1) disadvantage: there is always the danger of infection with surgery; (small) chance of damage to the cornea; only permanent if the shape of the eye doesn't change/lenses can be changed as the eye changes (1)

69. Correcting sight problems
Light from an object is bent/converged; by the lens; and cornea (or diagrams to show this); to form an image; on the retina; in short sight, the image forms in front of the retina; because the eyeball is too long; or the cornea/lens has too sharp a curve; in long sight, the image forms behind the retina; because the eyeball is too short; or the cornea/lens is not curved enough; spectacles and contact lenses make the image form on the retina; for short sight a diverging lens is needed; for long sight a converging lens is needed; contact lens are placed directly on the cornea; laser surgery changes the shape of the cornea; by making precise incisions/cuts in the cornea (6).

70. Different lenses
1 **a)** The point where the rays come together should be labelled as the focal point (1).
  b) 2.8 cm (1)
  c) It would be more sharply curved or its faces would have a smaller radius. (1) ('It would be fatter' is not acceptable.)
2 When rays of sunlight pass through a diverging lens they spread out/move apart (1); so the rays only appear to have come from the focal point/do not pass through the focal point (1).
3 **a)** D (1)
  b) power = 1/0.25 m (1) = 4 (1) dioptres
4 focal length = 1/power = 1/(3.75 dioptres) (1) = 0.266 (1) m

71. The lens equation
1 **a)** Rays from the lens meet (1) on the screen (1). (Or an answer describing how the lens makes rays from a point on the object meet at a point on the screen.)
  b) $1/f = 1/100\,cm + 1/25\,cm$ (1) = 0.010 + 0.040 = 0.050 (1) so $f = 1/0.050 = 20$ (1) cm
2 $1/u = 1/22\,mm - 1/25\,mm$ (1) = 0.0454 – 0.0400 = 0.0054 (1) so, $u = 1/0.0054 = 185$ (1) mm (You may have got an answer of 200 if you rounded $1/f$ to 0.045. You would still get the mark for this.)
3 **a)** $f = -40\,cm$, $u = 100\,cm$, $1/v = 1/f - 1/u = 1/-40\,cm - 1/100\,cm$ (1) = $-0.025 - 0.010 = -0.035$ (1) $f = 1/-0.035 = -28.60$ (1) cm (The minus sign must be in the answer to get the last mark.)
  b) virtual (1)

72. Reflection and refraction

1

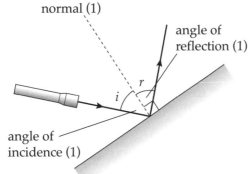

normal (1)

angle of reflection (1)

angle of incidence (1)

2 angle of reflection = angle of incidence = 90 − 50 (1) = 40° (1)
3 A ray travelling from the coin to the surface of the pool was bent away from the normal (1) (in the air) because the ray was refracted (1). (*'Because the speed of light changed/increased' is also correct.*)
4 The ray bends because the gamma rays/waves slow down (1) on entering the metal and the wavelength decreases (1). (*Or a labelled diagram showing this.*)
5 refractive index = 1.5, $\sin r = \sin i$/refractive index (1) = $\sin 40°/1.5 = 0.429$ (1) so, $r = 25.4°$ (1)

73. Critical angle

1 a) line, labelled A, showing ray in air bent away from the normal (1)
   b) line, labelled B, showing a ray reflected back into the glass with angle of reflection (roughly) equal to the angle of incidence (1)
2 Apparatus needed: source of a ray of light, protractor (1); How it is done: ray of light is sent into the glass block until it is just totally internally reflected off a surface (1); Measurements made: the critical angle/the angle between the ray and the normal where it is reflected, is measured (1).
3 a) $\sin c = 1/2.4 = 0.417$ (1), $c = 24.6°$ (1)
   b) refractive index = $1/\sin 39.12$ (1) = 1.585 (1)

74. Using reflection and refraction

1 1 mark for showing the ray bouncing off the inside of the fibre for its whole length; 1 mark for an attempt to show that the angles of incidence and reflection at each point are approximately equal.
2 a) At each point where the ray hits the side of the fibre the angle (of incidence) is greater than the critical angle (1), so the ray is totally internally reflected (1).
   b) Light is passed down an optical fibre to illuminate the inside of the body (1); reflected light passes up another optical fibre to the eyepiece (1). (*Your answer needs to explain how light gets into the body and how the observer sees the image.*)
3 A probe sends ultrasound waves into the body (1); at the boundary between different tissues the waves are reflected (1); the detector records the reflected waves. (1) (*Note: the waves may be partly reflected when being refracted across the boundary or totally internally reflected if they hit certain boundaries at greater than the critical angle*)
4 a) The ultrasound waves carry/transfer (1) energy (to the cancer cells) (1).
   b) The greater the intensity the greater the power/amount of energy (delivered to the cells) (1).
   c) If the beam widens the intensity falls, or to concentrate the energy on a small spot (1).

75. Physics extended writing 1

Answers can be found on page 129.

76. X-rays

1 a) CAEDB (all correct, 3 marks; 1 error, 2 marks; 2 errors, 1 mark)
   b) i) E (1)
      ii) A (1)
   c) With a higher charge on the anode, the electrons hit the anode with more energy (1); so the X-rays produced have a higher frequency (1). (*You could talk about higher ionization instead of higher energy*)
   d) The charged particles/electrons would hit/collide with particles/molecules of air (1); and would not reach the anode or no X-rays would be produced. (1)
   e) (In both) there is a flow (1) of electrons/charged particles (1).
2 Cathode temperature: control (1). Because the higher the temperature the more electrons are given off (1); so the more X-rays are produced (1).

77. X-ray calculations

1 a) $I = 3.125 \times 10^{18}$ per second $\times 1.6 \times 10^{-19}$ C (1) = 0.5 (1) A or amperes (1)
   b) KE = 1.6×10^{-19} C $\times 80 \times 10^3$ V (1) = 1.28×10^{-14} (1) J or joules (1)
2 a) number per second = I/q (1) = 0.4 A/1.6×10^{-19} C (1) = 2.5×10^{18} (1) electrons per second
   b) $v^2 = 1.6 \times 10^{-19}$ C $\times 40 \times 10^3$ V / $(\frac{1}{2} \times 9.11 \times 10^{-31}$ kg) (1)
      $= 1.4 \times 10^{16}$
      $v = \sqrt{(1.4 \times 10^{16})}$ (1) = 1.2×10^8 (1) m/s

78. Using X-rays

1 A (1)
2 When the source is close the X-rays will not spread out too much (attenuate) (1); so the intensity is high enough to give a sharp picture (1). *An alternative answer could be*: so that a more powerful X-ray source is not needed (1), with less risk then of damage to (surrounding) tissue/cells (1).
3 The benefits of using X-rays are greater than the risks of damage to tissues (1); precautions are taken to reduce the exposure to X-rays as much as possible (1). (*Answers such as 'using X-rays reduces the need for invasive exploratory surgery' would also be marked correct.*)
4 a) i) Any three from: use a CAT scan (1); the X-ray source is moved around the patient/head/brain (1); some of the X-rays are absorbed (1); detectors move on the opposite side (1); the X-rays detected build up into a picture of a slice through the patient (1).
      ii) 3 points from: Use a fluoroscope (1); X-rays from a source pass through the patient/heart (1); some of the X-rays are absorbed (1); some reach a screen/detector connected to a video camera (1); a TV/computer screen shows the (moving) pictures. (1)
   b) *The distance has doubled so (by the inverse square law) the intensity falls to $\frac{1}{4}$ of its value; $\frac{1}{4} \times 100$ W/m² (1) = 25 (1) W/m²*

79. ECGs and pacemakers

1 C (1)
2 a) a shorter time between traces/peaks or more traces/peaks on the screen/graph (1)
   b) The pacemaker helps the heart beat regularly/properly (1) by amplifying/transmitting the action potential (to the heart muscle) (1).
3 The oximeter compares amounts of red and infrared light (or the two colours) that are absorbed (1), because blood containing oxygen absorbs different amounts of each type of light compared with blood without oxygen (oxygenated blood absorbs less red light and more infrared light than unoxygenated blood) (1).
4 period = 0.9 s (1) (4.5 squares on graph); frequency = 1/0.9 s = 1.11 (1) Hz; heart rate = 1.11 Hz $\times$ 60 s = 66.6 beats per minute (67 would also be accepted) (1)

80. Properties of radiation

1 a) alpha – particles made up of two protons and two neutrons (1); beta – fast-moving electrons (1); positron – particles with positive charge and same mass as electron (1); gamma – electromagnetic radiation (1); neutron – particles with same mass as proton but no charge (1)
   b) gamma – beta – alpha (1)
2 a) The number of protons and electrons is the same (1); and protons and electrons have equal and opposite charges (1).
   b) The electrons are not counted or only the protons and neutrons are included (in the nucleon number) (1); because the mass of the electron is negligible/very small/very much less than a proton/a 2000th of a proton (1).
3 a) A neutron (1) in the nucleus becomes a proton (1) and a beta particle.
   b) (1 mark for each correct row)

Element	Proton number	Nucleon number	Radiation emitted	New proton number	New nucleon number
lithium	3	8	beta	4	8
uranium	92	238	alpha	90	234
americium	95	239	gamma	95/no change	239/no change

81. Balancing nuclear equations

1 $\beta-$ decay: electrons emitted when a neutron changes into a proton (1); $\beta+$ decay: positrons emitted when a proton changes into a neutron (1)
2 a) alpha or helium nucleus or $^4_2$He (1)
   b) beta $+$/$\beta+$ or positron or $^0_1$e (1)
3 a) top 131 (1); bottom 54 (1)
   b) top 235 (1); bottom 92 (1)
   c) top 12 (1); bottom 4 (1)
4 a) $^{99}_{42}$Mo $\rightarrow$ $^{99}_{43}$Tc $+$ $^0_{-1}$e (1)   b) beta particles (1)

82. The nuclear stability curve

1 a) D (1)
   b) 50 or 51 (1)
2 a) the number of protons in radium-224 is greater than 82 (1); so it undergoes alpha decay (1)
   b) oxygen-13 undergoes $\beta+$ decay/gives out a positron (1) because it has less than the stable number of neutrons/is below the stable line (1).
3 Any three from: bismuth-212 has more neutrons than the most stable isotope (1); bismuth-212 has more than 82 protons (1); in both forms of decay the nucleus moves closer to the line of stable isotopes (1); alpha decay moves the nucleus diagonally down and leftwards/reduces the number of protons and neutrons (1); $\beta-$ decay moves the nucleus to the right/reduces the number of neutrons (1).

83. Quarks

1 a) B (1)
   b) A neutron is made up of two down (1) quarks, each with a charge of $-1/3$ e (1); and one up (1) quark with a charge of $+2/3$ e (1); making a total of zero.
   c) All quarks have a mass of $1/3$ (1); protons and neutrons contain 3 (1) quarks giving a total of 1.
2 a) A quark has changed from down to up (1).
   b) beta/$\beta-$ or electron emission (1)
3 In a proton (1) an up quark changes to a down quark (1); a charge of +1 is given to/removed by/taken by the positron (1).

84. Dangers of ionising radiation

1 a) Any three from: the dose of ionising radiation given to the patient is kept as low as possible (1); they stay as far from the radioactive source as possible (1); the source is shielded (lead glass windows, lead aprons worn); the source is contained (lead lined boxes). *You could also say that the time patients are exposed to radiation is also limited* (1).
   b) Medical workers may be exposed to ionising radiation at any time (1) so it is important to know how much radiation they receive. (1).
2 a) ionising radiation kills many cells (1) causing damage to tissues/organs (1). *(Saying that cells die is not enough as people only become sick when tissues and organs fail.)*
   b) Ionising radiation can damage DNA (in cells) (1) causing mutations (that may lead to cancer) (1).
   c) Cells in unborn babies can be damaged or killed or mutations can occur in sex cells/sperm/eggs (1); causing the foetus to develop incorrectly (1).
   d) The effects of radiation sickness /high dosage appear quickly (1); but cancers can take many years to develop (1); OR people with radiation sickness fall ill/die quickly (1); but not everyone who has/will get cancer will have died yet (1); OR there are other causes of cancer (1); which make it difficult to identify who has died as a result of the accident (1).

85. Radiation in hospitals

1 CBDEA (all correct, 3 marks; 4 in correct order, 2 marks; 2 or 3 in correct order, 1 mark)
2 a) B/I-131 (1); because the β rays/particles can travel a few centimetres to kill the cells nearby (1). *(The answer should suggest that the cells that are destroyed are close to where the tracer is.)*
   b) A/Co-60 (1); because the gamma rays can travel into the body some distance to reach the diseased cells/tissue (1).
3 a) If the patient has cancer then the cells must be destroyed, which requires a fairly high dose (1); in diagnosis the patient may be healthy so as few cells must be damaged as possible (1). *(Answers will be accepted that point out that cancerous cells are targeted in patients and damage to healthy cells is limited.)*
   b) Palliative care means relieving pain (1); this can be very important to improve the quality of life of patients even if healthy cells are damaged (1).

86. Radioactivity summary

1 a) mass: 1/2000 (1); charge: -1 (1)
   b) a neutron (1) turns into a proton and electron (1) *(You could also answer this question in terms of quarks.)*
   c) The beta rays are stopped/absorbed/cannot pass through the metal box (1); so the medical workers are protected from the harm that the beta particle can cause (1).
2 a) C (1)
   b) top: 18 (1); bottom: 8 (1)
   c) an up (1) quark turns into a down (1) quark
   d) (Because of the short half-life) the amount of (fluorine-18) in the sample falls quickly (1); so if it is made just before it is used there will be enough to produce a clear image (1). *(Answers will be accepted describing the loss of fluorine-18 if it has to be transported over a large distance.)*

87 and 88. Physics extended writing 2 and 3

Answers can be found on page 130.

89. Particle accelerators

1 C (1)
2 a) high frequency alternating voltage (1)
   b) electromagnet (1)
3 Protons/charged particles (1); from the cyclotron collide with a target of a stable isotope (1).
4 Any two from: build a picture of sub-atomic particle (1); find new (fundamental) particles (1); find the properties of particles (1); test theories/models/answer questions about how the universe formed/how atoms/matter is constructed (1); find the Higgs particle (1). *(Any answer that links what the LHC does to the problems scientists want to solve gets a mark. Many scientists can share data/ideas.)*

90. Collisions

1 D (1)
2 The second truck moves with the same kinetic energy and velocity (1); as the first truck had because energy and momentum are conserved/not changed (1).
3 a) It is moving in the opposite direction (1) from the trolley on the left (1).
   b) The kinetic energy the trolleys had before the collision is lost/not conserved (1); it is changed into heat/sound (1).
   c) Momentum does not change/is conserved (1); the total momentum is zero or the momentum of the right trolley is equal and opposite/negative to the left trolley (1). *(The answer should show an understanding that the total momentum before and after the collision is zero because momentum is a vector quantity.)*
   d) Inelastic (1); because the kinetic energy was not conserved (1).

91. Calculations in collisions

1 a) total momentum before collision = momentum of small ball $= m \times v = 0.25 \text{ kg} \times 4 \text{ m/s} = 1$ (1) kg m/s; total momentum after collision $= 1$ (1) kg m/s; velocity of larger ball $=$ momentum/mass $= 1 \text{ kg m/s}/0.5 \text{ kg} = 2$ (1) m/s
   b) total KE before collision $= \frac{1}{2} \times 0.25 \times 4^2 = 2$ J (1); total KE after collision $= \frac{1}{2} \times 0.5 \times 2^2 = 1$ J (1); KE not conserved (1) *(You may be awarded the last mark if your calculation of either or both of the kinetic energies was incorrect.)*
2 momentum before collision $= 1 \times 1 \times 10^8 = 1 \times 10^8$ (1) = momentum after collision; so velocity of beryllium $= 1 \times 10^8/8$ (1) $= 0.125 \times 10^8$ or 1.25×10^7 (1) m/s
3 Total momentum before collision $= 1 \times 4 + 2 \times -2 = 0$ kg m/s (1); total KE before collision $= \frac{1}{2} \times 1 \times 4^2 + \frac{1}{2} \times 2 \times 2^2 = 12$ J (1); for elastic collision both momentum and KE are conserved; momentum after $= 1 \times v_1 + 2 \times v_2 = 0$, so $v_1 = 2 v_2$ (1); KE after $= \frac{1}{2} \times 1 \times v_1^2 + \frac{1}{2} \times 2 \times v_2^2 = 12$; $\frac{1}{2} \times (2v_2)^2 + v_2^2 = 3v_2^2 = 12$, so $v_2 = +/- 2$ m/s and $v_1 = -/+ 4$ m/s (1); i.e. the balls have the same speed as before the collision but probably in the opposite direction (1).

92. Electron–positron annihilation

1 a) A (1)   b) B (1)
2 a) The mass of the positron and electron is converted into the energy (of the gamma rays) (1); so mass-energy is conserved/stays the same (before and after the collision) (1).
   b) The energy of the gamma rays has a mass equivalent/momentum (1); and as they are emitted in opposite directions their total momentum is zero (1); which is the same as the initial total momentum of the positron and electron (1).
3 total energy released $= 2 \times 9 \times 10^{-31}$ kg $\times (3 \times 10^8$ m/s$)^2$ (1) $= 1.62$ (1) $\times 10^{-13}$ (1) J or 162 (1) $\times 10^{-15}$ (1) J

93. Kinetic theory

1 a) Solids: Particles can vibrate but cannot move freely, and Particles are held closely by strong forces; Liquids: Particles are close together but can move past each other; Gases: Particles are far apart and move quickly (all 4 correct, 3 marks; 3 correct, 2 marks; 2 correct, 1 mark)
   b) The particles in the gas collide (1) with (the surface of) the balloon (1)
2 a) absolute zero (1)
   b) The particles move more slowly (as the temperature falls) (1); and stop moving at absolute zero (1).
3 a) temperature in K $= -78 + 273$ (1) $= 195$ (1) K
   b) The graph is a straight line (with a positive gradient) (1) through the origin/0 K (1). *(Stating that the graph shows that the average KE is directly proportional to the Kelvin temperature is awarded 2 marks.)*

94. Ideal gas equations 1

1 a) C (1)
   b) $T_1 = 77°C = 350$ K, $T_2 = 27°C = 300$ K (1), $V_2 = 0.6$ cm^3
       $V_1 = V_2 T_1 / T_2 = 0.6$ cm$^3 \times 350$ K / 300 K (1) = 0.7 (1) cm^3

2 a) The cylinder holds gas that has been compressed to a fraction of its normal volume (1) by very high pressure (1).
   b) $1 \times V_1 = 10$ l $\times 200$ atm, $V_1 = 2000$ l (1); time that the gas will last = 2000 l/2 l/min = 1000 min (1); 16 hours is $16 \times 60 = 960$ min (1); so the cylinder will last at least 16 hours (even allowing for the gas left in the cylinder).

95. Ideal gas equations 2

1 a) $P_1 = P_2 V_2 / V_1$ (1) = $640 \times 100\,000$ Pa/8 l (1) = 8 000 000 or 8×10^6 (1) Pa
   b) time = total volume/rate = 640 l/2 l/min = 320 min (1); but 8 l will be left in the cylinder at atmospheric pressure, so actual supply will only last 316 min (1). (*2 marks for 316 min with working.*)

2 a) $P_1 = 100\,000$ Pa, $V_1 = 0.0005$ m^3, $T_1 = 17°C = 290$ K; $P_2 = 34\,000$ Pa, $T_2 = -26°C = 247$ K

   $V_2 = \dfrac{P_1 V_1 \times T_2}{T_1 \times P_2}$ (1) $= \dfrac{100\,000 \text{ Pa} \times 0.0005 \text{ m}^3 \times 247 \text{ K}}{290 \text{ K} \times 34\,000 \text{ Pa}}$ (1)

   $= 0.0013$ cm^3 (1)

   b) The climbers need to top up the oxygen they breathe (1); because (the volume of air has more than doubled) the climbers can only take less than half/two-fifths of the air they need on each breath (1).

3 $T_2 = \dfrac{P_2 V_2 T_1}{P_1 V_1}$ (1) $= \dfrac{104\,000 \times 2800 \times 285}{101\,000 \times 2400}$ (1) = 342 (1) K (69°C)

96 and 97. Physics extended writing 4 and 5

Answers can be found on page 130.

Extended writing answers

Below you will find a list that will help you to check how well you have answered each Extended Writing question. A full answer will contain most of the points listed but does not have to include all of them and may include other valid statements. Your actual answer should be written in complete sentences. It should contain lots of detail and link the points into a logical order. You are more likely to be awarded a higher mark if you use correct scientific language and are careful with your spelling and grammar.

Biology extended writing 1

Photoperiodicity is the response of plants; to changes in day length/length of night; stimulates seed germination; stimulates growth; at the right time/time quoted (e.g. spring); idea of increasing/maximising photosynthesis; so more sugar is made; stimulates flowering; synchronised to other flowers (of same species); to increase chance of pollination; synchronised to pollinator presence/activity.

Biology extended writing 2

Boy had not previously had cowpox; Jenner infected boy with cowpox; took cowpox pus from milkmaid and gave it to boy; subsequently tried to infect boy with smallpox; cowpox antigen caused primary immune response to occur; memory lymphocytes to cowpox antigen formed; smallpox antigen very similar/similar shape to cowpox antigen; exposure to smallpox antigen stimulated memory lymphocytes to produce antibodies; rapidly; in great quantities; antibodies aided in destroying smallpox/description of one form of antibody action; reference to secondary immune response.

Biology extended writing 3

Climate became colder; adaptation of clothing, home and equipment to suit new environments; water froze in glaciers/ice caps; sea levels were lowered; more dry land; less water/sea to cross; correct reference to example, e.g. Ice Age, less water/distance between Africa and Asia; therefore there was a migration of people to new areas; more stable climate led to people settling down and starting to grow crops. (You would also get credit if you made the link between a warmer climate and the fact that people could move away from the equator.)

Biology extended writing 4

High in (essential) amino acids/protein; for efficient growth; low in cholesterol/fats/saturated fats; which reduces likelihood of cardiovascular disorders/condition described; reduces likelihood of obesity; low in sodium; so less likely to cause high blood pressure; high in fibre; so less likelihood of bowel cancer/obesity/reduces gut transit time; comparison to meat such as lower in saturated fat than meat; and higher in fibre than meat.

Biology extended writing 5

DNA is modified by addition of a gene; from another species; gene cut out from original DNA using restriction enzyme; this creates sticky ends; plasmid is opened using the same restriction enzyme; ligase is used; to join the complementary sticky ends together; to produce a modified/recombinant plasmid; this plasmid inserted into a different/new bacterium.

Chemistry extended writing 1

Qualitative: shows which ions are present; for aluminium ion, add sodium hydroxide solution; white precipitate forms; of aluminium hydroxide; $Al^{3+} + 3OH^- \rightarrow Al(OH)_3$; this precipitate dissolves when more sodium hydroxide is added; to form colourless solution; to test for sulfate ion add nitric acid (or hydrochloric acid); and barium chloride; white precipitate forms; of barium sulfate; $Ba^{2+} + SO_4^{2-} \rightarrow BaSO_4$.

Quantitative: known volume of water; precipitate out barium sulfate; filter; weigh the precipitate; work out number of moles of (aluminium) sulfate in the solution.

Chemistry extended writing 2

Copper (II) chloride: copper oxide/hydroxide/carbonate; which is an insoluble base; add it in small amounts; to hydrochloric acid; it reacts with acid to form soluble salt; stops reacting when acid is all used up; and so stops dissolving; so no indicator needed; excess base in beaker shows reaction is over; and can be filtered off.

Sodium chloride: hydrochloric acid; sodium oxide/sodium hydroxide/sodium carbonate; these are soluble bases; need an indicator; to tell when acid is neutralised; such as methyl orange/litmus/phenolphthalein; colour change given; indicator needed because you can keeping adding base and it will dissolve; and solution goes through neutral to alkaline; this is called a titration; can work out the relative volumes of alkali and acid needed for full neutralisation; and repeat with no indicator; to give pure salt.

Chemistry extended writing 3

Electrolyte; is a copper salt, such as copper (II) sulfate; copper ions; migrate to negative electrode; which is the cathode; turn into copper atoms; layers of pure copper build up on electrode; $Cu^{2+} + 2e \rightarrow Cu$; gain of electrons; reduction; impure copper contains copper atoms; at the positive electrode/anode; these atoms ionise; and dissolve into the electrolyte; $Cu \rightarrow Cu^{2+} + 2e$; loss of electrons; oxidation; impurities in the block at the positive electrode; do not dissolve/fall to bottom of cell; to form a sludge.

Chemistry extended writing 4

Fertilisers contain nitrogen and often other elements; such as potassium/phosphorus; which increase the growth; of crop plants; this is beneficial; as the world population increases; and more people need to be fed; however, over-use of fertilisers can cause problems; fertilisers are soluble; so rain water can wash them through the soil; in a process called leaching; build up in ground water; and get into streams/rivers/lakes; where they encourage growth of algae and plants; this removes oxygen from the water; and leads to death of other river organisms; eutrophication; concluding personal opinion on whether fertilisers are positive or negative.

Chemistry extended writing 5

Oil/fat molecules are esters; which can be broken down/hydrolysed; by boiling; with a concentrated solution; of an alkali; the long hydrocarbon chain; comes from the original fat molecule; and the alkaline solution turns them into salts; of carboxylic acids; which are soluble; hydrocarbon 'tails'; dissolve in/interact with; grease molecules; they are hydrophobic; ionic 'heads'; are hydrophilic; and line up pointing out of grease molecule; surrounding grease molecule and helping 'pull' it off clothes/hands; and into solution.

Physics extended writing 1

When light is incident on a boundary between two media (such as glass and air) it can be reflected or refracted; when light is reflected the angle of incidence equals the angle of reflection; light travels at different speeds in different media; when light travels out of a medium where it travels more slowly than in air (e.g. glass); it is refracted away from the normal; if the angle the light makes with the normal in the glass is greater than the critical angle all the light is reflected (back into the glass); this is called total internal reflection; when light travels down an optical fibre it always hits the side of the fibre at an angle greater than the critical angle; so the light travels along the fibre and cannot escape from it, even if the fibre is bent; in communications pulses of light travel along the optical fibres; in endoscopes light is sent down some optical fibres to illuminate the inside of the body/fallen building; light is reflected up the optical fibres; and is focused by an eyepiece lens.

Physics extended writing 2

In a CAT scanner the X-ray tube rotates around the patient; X-rays passing through the patient reach detectors; which build up an image; the image is 3D (3-dimensional); which is viewed on a computer screen; X-rays are absorbed by the organs of the body; different organs absorb different amounts; dense/bone tissues absorb the most; precautions – the X-rays are produced in a narrow beam; shielding stops the X-rays spreading; the operators are in another room/a long way from the scanner; so are protected because the doubling of distance reduces intensity of the X-rays to a quarter (the inverse square law); the value of the CAT scan in diagnosing illness is greater than the risk of harm.

Physics extended writing 3

Radioactive decay produces alpha particles; which are helium nuclei; made up of two protons and two neutrons; positively charged particles; which are slow moving; and very ionising; $\beta-$ (beta minus) particles; which are fast moving electrons; which have a negative charge; formed by a neutron becoming a proton and an electron; $\beta+$ (beta plus) particles; which are fast moving positrons; positive charge; low to medium ionisation; gamma rays; which are a form of electromagnetic radiation; formed when the particles in the nucleus rearrange and release energy; no mass; most penetrating; neutrons; which have no charge; neutrons are as penetrating as gamma rays; $\beta+$ formed by proton; changing into a neutron and positron; type of radiation depends on position of radioisotope in relation to nuclear stability curve; for Z. 82 alpha most likely; for Z, 82: above the curve, too many neutrons – $\beta-$ decay; below the curve, too few neutrons – $\beta+$ decay; in beta decay one quark changes into another: in $\beta-$ radiation a down quark changes to an up; in $\beta+$ radiation an up quark changes to down.

Physics extended writing 4

Particle (e.g. electron) emitted by radioactive source enters centre of cyclotron; particle is accelerated by electric field; as it crosses gap between the D-shaped electrodes; magnetic field bends path of particle in a circle; particle moves in a circle because (magnetic/centripetal) force on charged particle acts at right angles to motion of particle/towards centre circle; when particle leaves cyclotron it moves in a straight line/at a tangent to the circle; because there is no longer a (centripetal) force acting; the particle has a large momentum due to its speed (close to the speed of light); the particle bombards the target; causes nuclear reactions in the target; these changes produce useful radioisotopes; produced on demand; close to where they are needed; so that they still have sufficient activity when used if their half-life is short; less risk that ionising radiation can escape.

Physics extended writing 5

The kinetic theory states that everything is made of particles; in solids and liquids the particles are close together; held by strong forces; in a gas the particles are far apart; move around quickly; move to all parts of the container; the temperature of a gas is a measure of the average kinetic energy of the particles; the faster the particles move the higher the temperature of the gas; the pressure of a gas is caused by the force exerted when particles of a gas hit the walls of the container; the faster the particles move the more collisions there will be; the more force there is in each collision; so, the higher the temperature, the higher the pressure; experiments show that this prediction is correct; a graph of temperature against pressure is a straight line; pressure is proportional to temperature of a gas; the pressure is zero when the temperature is at absolute zero, 0 K or $-273°C$; absolute zero is when the particles have no kinetic energy/are not moving; so they cannot exert a pressure; absolute zero cannot be measured directly because gases change into liquids before they are cooled to absolute zero; the laws only apply to gases.

Biology practice exam paper

1 a) i) carbon dioxide/water (1)
    ii) Urea is made in the liver (1) from the breakdown of excess amino acids (1).
  b) i) X = in the indent of the Bowman's capsule (1); Y = on the descending/ascending or the actual loop (1)

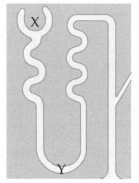

    ii) One from: Less glucose in the blood (1); some glucose in urine. (1) *(You do not get the mark if you say 'more' glucose in the urine, as normally the urine does not contain any glucose at all.)*
  c) less ADH produced (1); so less water reabsorbed in the kidneys/nephron/collecting duct (1); so less water in the blood (1)
  d) ureter to bladder to urethra (1 for correct parts named, 1 for correct sequence)

2 a)

Description	Appropriate cell
Cell produced antibody in person P	Lymphocyte (1)
Cell that produces antigen	Bacterium/pathogen (1)
Cell that makes person Q immune	Memory lymphocyte (1)

  b) swelling at point of injection/mild symptoms/(rare) allergic reaction (1)
  c) monoclonal antibody (1)
  d) specific to antigen (1); (this) antigen is only found on cancer cells (1); radiotherapy is not specific/targets various (other) rapidly dividing cells as well/such as cells that produce hair (1)

3 a) i) constant temperature (1)
    ii) optimum temperature/effective temperature for bacterial growth (1)
    iii) (live) bacteria in pasteurised milk (1); grow rapidly at 36°C (1); no live bacteria/bacteria dead in sterilised milk (1)
    iv) To show that milk does not cause resazurin to change colour/ it is the different types of milk/ process that causes the colour change/as a comparison (1)
  b) i) The number of bacteria rises slowly at first (1) and then increases very fast/shows exponential growth (1).
    ii) The growth will slow down (1) because the bacteria will run out of food/may produce toxins which inhibit growth (1).

4 a) i) thorns that can be hollowed out (1); food supplied (1)
    ii) Protect from elephants/herbivores eating it (1); so does not reduce photosynthesis/so photosynthesis not reduced (1)
  b) new flowers are to attract pollinators/bees (1); stops ants from competing with pollinators/reducing pollen/ nectar/eating the pollen/nectar (1)
  c) Old flowers have already been pollinated (1); so don't need remaining pollen/nectar (1)

5 a) to swim towards/to an egg (1); to fertilise it (1); by delivering its haploid nucleus/half the genetic material/23 chromosomes (1)
  b) egg cell membrane changes (1)
  c) colour blindness is sex linked; the gene for being able to tell red and green apart is carried on the X chromosome; girls have two XX chromosomes and boys have an X

and a Y; the gene for being able to tell red and green is dominant; but there is a recessive allele which means that you cannot tell red and green apart; if a girl is to be colour blind she has to inherit a recessive allele from *both* parents which is rare; if a boy inherits one recessive allele he will be red/green colour blind; Punnett Square correctly drawn; and correctly filled in; if the mother is a carrier the probability of boy being red/green colour blind is 50% (6).

6 a) use enzymes (1); any two from: protease to break down protein stains; lipase to break down lipid/grease/fat stains; carbohydrase to break down starch stains (2)
  b) Four from: To make lactose-free milk (1); by converting/ breaking down the lactose (1); into other sugars/named sugar (1); to keep the lactase/enzyme attached (1); so does not contaminate product/easier to extract/reuse. (1)
  c) More rapid growth; easier to manipulate/handle/ contain; possible to grow in optimal conditions; grown in any place/independent of external environment/climate; can use waste products/example given; therefore cheaper; better for the environment; *Fusarium* makes mycoprotein; low in saturated fat; no cholesterol (6).

Chemistry practice exam paper

1 a) C (1)
  b) i) propene (1)
    ii) carbon skeleton: C – C = C or C = C – C (1); all other hydrogens shown $CH_3 - CH = CH_2$

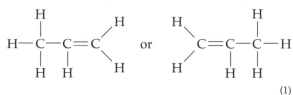

(1)

  c) Any two from: differ from each other by CH_2 (1); have the same chemical properties (1); have gradual change in physical properties. (1)
  d) hydrogen (1); in the presence of a catalyst/adds across the C = C double bond (1)

2 a) B (1)
  b) plotting of at least three points correctly = (1); plotting all points correctly = (2); line of best fit = (1)

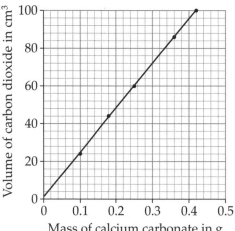

  c) i) line drawn up from 0.3 g of calcium carbonate to hit graph, then taken horizontally to y-axis, giving a volume of 72–74 cm³ (the exact value will depend on the graph drawn) (1)
    ii) answer from i) ÷ 0.30 (1) × 100 (1) [= 24 000 cm³]
    iii) molar volume/volume of 1 mole of any gas (1)

3 a) i) A (1)
    ii) sodium hydroxide (solution) (1); green precipitate (1)
  b) i) silver iodide (1)
    ii) $AgNO_3 + KI \rightarrow AgI + KNO_3$ (correct formula of silver iodide, 1; rest of equation, 1)

iii) Any two from: removes other ions (1); that would interfere/also give a positive result (1); such as the carbonate ion (1).

c) Any two from: salt also contains chloride ions (1); in much larger quantity/iodide ions are only present in small amounts (1); so precipitate more likely to be white/off-white (1).

4 a) i) nitric acid (1)

ii) All acid is used up/neutralised (1); and zinc oxide is not soluble/some zinc oxide remains unreacted. (1)

iii) C (1)

b) i) EITHER: add indicator to sodium hydroxide solution in conical flask (1); add acid until it changes colour (1) (*You can score the first mark for naming a correct indicator, such as methyl orange, litmus or phenolphthalein. However, you will not get a mark for universal indicator, as this indicator is not used for titrations.*); OR: take a very small sample of solution from conical flask and place onto indicator/litmus paper (1); until neutral/colour change (1)

ii) $H_2SO_4 + 2\,NaOH \rightarrow Na_2SO_4 + 2\,H_2O$ (1)

iii) number of moles of NaOH = 25 cm^3/1000 × 0.5 mol dm^{-3} = 0.0125 moles (1); number of moles of H_2SO_4 = 0.0125 moles/2 = 0.00625 moles (1); volume of H_2SO_4 = 0.00625 moles × 1000/0.2 mol/dm^3 = 31.25 cm^3 (1)

5 a) i) Any two from: lead bromide melts (1); ions become free to move (1); and so can conduct (DC) electricity/complete the circuit. (1)

ii) A (1)

iii) Electrons are gained therefore this is a reduction (1); $Pb^{2+} + 2e \rightarrow Pb$ (correct formula for lead ion, 1; equation correct, 1)

b) Metal fork is electroplated; with silver; positively charged silver ions; in the electrolyte; move to negative electrode; gain electrons; to form silver; $Ag^+ + e \rightarrow Ag$; silver block at positive electrode gets smaller; silver atoms; turn into silver ions; by losing electrons; $Ag \rightarrow Ag^+ + e$; may see some bubbles at negative electrode; some hydrogen produced; from hydrogen ions; formed as water molecules split up; $2H^+ + 2e \rightarrow H_2$ (6).

6 a) i) C (1)

ii) ethanoic acid (1)

b) i) ethyl ethanoate (1); $CH_3COOC_2H_5$ (1)

ii) use to make flavourings (1); or solvents (1)

c) Fermentation; from glucose/sugar; using yeast; at a warm temperature; anaerobic process; $C_6H_{12}O_6 \rightarrow 2C_2H_5OH + 2CO_2$; useful for countries that grow lots of sugar beet; but gives a poor yield (about 15%); so ethanol needs to be concentrated; by fractional distillation; but sugar beet is renewable; hydration of ethene; gives purer product; and is faster; but needs ethene from crude oil; so only useful for countries with oil reserves; oil is non-renewable; $C_2H_4 + H_2O \rightarrow C_2H_5OH$; needs high temperature; high pressure; and a catalyst (6).

Physics practice exam paper

1 a) C (1)

b) i) Kinetic energy is not conserved (1); (some) is changed to heat/sound. (1)

ii) The total momentum of the two balls before and after the collision is zero (1); because they move at the same speed in opposite direction or their velocities are the same size but in opposite directions. (1)

c) $v^2 = 2 \times KE/m = 2 \times 0.024$ J/0.012 kg = 4 (1) $v = 2$ (1) m/s; momentum = 0.012 kg × 2 m/s = 0.024 (1) kg m/s.

2 a) A (1)

b) period (from graph) = 1.2 (1) s; frequency = 1/period = 1/1.2 = 0.8333 beats per second; which is 0.8333 × 60 = 50 (1) beats per minute

c) heart rate higher or lower than normal (1); the shape of the graph/PQRST pattern (1)

d) The pacemaker detects the action potentials (1); and amplifies them. (1) (You could also say the pacemaker detects if there is an irregular beat and sends a pulse to make the heart contract.)

e) the amount of oxygen carried by the blood (1)

3 a) i) internal examination (1)

ii) Light rays hit the side of the optical fibre at an angle greater than the critical angle (1); and so are totally internally reflected (1).

b) i) The speed of the light rays/waves is slower in the lens (1); so they are bent as they cross the boundary between glass and air (1). (Your answer must make clear that the bending takes place at the boundary and not inside the lens.)

ii) power = 1/f = 1/0.15 m (1) = 6.66 (1) dioptres

iii) $1/v = 1/f - 1/u$ (1) = 1/0.15 m − 1/0.08 (1) v = −0.584 (1) m (The answer must have the minus sign.)

4 a) stops air entering the tube – glass tube; thermionic emission of electrons – heated cathode; prevents X-rays escaping except through the window – lead lining (all correct – 2 marks, 2 correct – 1 mark)

b) The electrons must not collide with air/gas (molecules) (1); because this will stop the electrons reaching the target/anode. (1)

c) to attract the electrons to the anode/target (1); at high speed (1)

d) i) KE = eV = 1.6 × 10^{-19} C × 120 × 10^3 V (1) = 1.92 × 10^{-14} (1) J

ii) $N = I/q$ = 6.4 × 10^{-5} A /1.6 × 10^{-19} C (1) = 4 × 10^{14} (1)

5 a) neutron mass 1 (1), charge 0 (1)

b) It reduces its mass/mass number (1); and charge/number of protons/atomic number. (1)

c) top 188 (1); bottom 80 (1)

d) Alpha particles/rays travel (less than) a millimetre/very short distance through tissue; beta particles will travel several mm/further; gamma will travel though the whole body/furthest; alpha are most ionising, beta next, gamma least; so, alpha most effective at killing cancer cells; but in TAT (once they have reached the cells) will not harm healthy surrounding cells; beta and gamma kill healthy cells around/on their way to the tumours; so, sources used in TAT can be less powerful than in other methods; less harmful to medical staff handling the radioactive materials (6).

6 a) 0.75 (1)

b) point plotted accurately (1); best fit straight line (1); through origin (1)

c) As the volume decreases more particles hit a unit area in a unit time (1) increasing the pressure (1).

d) A full answer will include most of the following points but does not have to include all of them: All the points do not fit on the best fit straight line; the point at 20cm^3 in particular is anomalous. Best fit line does not pass through the origin; perhaps points not calculated correctly. Volume scale not read accurately; pressure gauge not read accurately; pressure gauge faulty; the apparatus leaked/air lost from tube; so mass of gas not constant; temperature varied/not constant; not enough readings at different pressures taken; readings not repeated (6).